CONSCIOUSNESS AND THE SOURCE OF REALITY

OTHER BOOKS BY ROBERT G. JAHN AND BRENDA J. DUNNE

Margins of Reality:
The Role of Consciousness in the Physical World

Filters and Reflections:
Perspectives on Reality
(with Zachary Jones and Elissa Hoeger)

BY ROBERT G. JAHN

The Role of Consciousness in the Physical World
AAAS Selected Symposium: 57

Physics of Electric Propulsion

CONSCIOUSNESS AND THE SOURCE OF REALITY

The PEAR Odyssey

Robert G. Jahn *and* Brenda J. Dunne

ICRL Press
Princeton, New Jersey

Consciousness and the Source of Reality
Copyright © 2011 by Robert G. Jahn and Brenda J. Dunne
ISBN: 1-936033-03-8

Cover: *BUNKAN:* aurora borealis image by Odd Erik Garcia

All rights reserved, including the right to reproduce this book or portions thereof in any form whatsoever. For information address:
ICRL Press
211 N. Harrison St., Suite C
Princeton, NJ 08540-3530

CONTENTS

TO THOSE WHO DARE...

To live their dreams,

To dance with uncertainty,

To embrace the Source.

PROLOGUE:
THE PEAR EXPERIENCE

When you set out for Ithaka
ask that your way be long,
full of adventure, full of instruction.
The Laistrygonians and the Cyclops,
angry Poseidon —do not fear them:
such as these you will never find
as long as your thought is lofty, as long as a rare
emotion touch your spirit and your body.
— C. P. Cavafy[1]

It started as an unused storage area next to the machine shop in the basement of Princeton University's School of Engineering and Applied Science. In an almost organic manner, it soon doubled in size and then progressed to four, and then to five small rooms, and eventually evolved into the unique and dynamic environment that was the Princeton Engineering Anomalies Research laboratory. It gradually became furnished with sophisticated high-tech equipment, environmental sensors and monitors, chart recorders, oak paneling, carpeting, and a large orange velvet sectional sofa facing an enormous 10' × 6' "pinball machine" that occupied an entire wall. Framed NASA images of planets and galaxies adorned the walls, along with countless cartoons, and a "McArthur's Corrective Map of the World" that offered a novel geographical perspective by having South at the top. Assorted notices were posted on the doors, one of which advised the staff "Don't Speculate, Concatenate;" another instructed that "There shall be tea and cookies, and juice in tiny little cans." A sign on the door to one of the experiment rooms read "Operator at Play," in an assortment of languages. Mismatched metal file cabinets and desks rescued from the university's surplus warehouse lined every available wall of the two staff offices, and two comfortable reclining chairs that arrived bearing tags reading "Hello, I'm Comforto the Incredible" were provided for the experimental operators. Overseeing it all was a small plush frog with a silly grin who became the official mascot of the laboratory, along with a virtual Noah's Ark of other stuffed animals.

The activities that took place in this exceptional space were as eclectic as its fixtures. Staff discussions ranged from the relative merits of Bayesian vs. frequentist statistics, the dynamics of laminar and turbulent fluid flows, quantum entanglement, and the proper calculation of *a priori* probabilities for remote perception descriptors, to the functions of the unconscious mind, traditions of Eastern philosophy, indigenous healing techniques, and the nature of reality. The blackboard shared an assortment of mathematical equations with "To Do" lists and pithy quotations. The permanent staff, comprising an exceptional combination of professional and personal backgrounds and perspectives, was regularly augmented by a variety of shorter-term specialists and interns.

This was the PEAR lab.

When we first embarked on this exotic scholarly journey more than three decades ago, our aspirations were little higher than to attempt replication of some previously asserted anomalous results that might conceivably impact future engineering practice, either negatively or positively, and to pursue those ramifications to some appropriate extent. But as we followed that tortuous research path deeper into its metaphysical forest, it became clear that far more fundamental epistemological issues were at stake, and far stranger phenomenological creatures were on the prowl, than we had originally envisaged, and that a substantially broader range of intellectual and cultural perspectives would be required to pursue that trek productively. This text is our attempt to record some of the tactics developed, experiences encountered, and understanding acquired on this mist-shrouded exploration, in the hope that their preservation in this format will encourage and enable deeper future scholarly penetrations into the ultimate Source of Reality.

Reference

---

[1] Constantine Petrou Cavafy. "Ithaka." In George Savidis, ed., trans. Edmund Keeley and Philip Sherrard, *Collected Poems*. Revised Edition. Princeton, NJ: Princeton University Press, 1992.

INTRODUCTORY NOTES

PEAR was conceived and implemented in the late 1970s for the primary purpose of determining the potential vulnerability of physical systems and technological processes involving random elements to the conscious or unconscious intentions of their human operators. Its ancillary goal was an attempt to comprehend the implications of any such anomalous interactions for a broader understanding of human consciousness and its role in the establishment of physical reality. Experimental results over the first decade of a coordinated menu of empirical studies and complementary theoretical models validated the original technical concerns via rigorous demonstrations of an array of anomalous phenomena that had been widely proclaimed over several millennia of anecdotal reportage and featured in countless books and professional journals. But even deeper issues were to emerge.

These early PEAR results resulted in well over 150 articles, technical reports, and chapters in scholarly books. Many of these were published in conference proceedings and in a variety of refereed scientific journals—in particular the *Journal of Scientific Exploration*, published by the Society for Scientific Exploration—and were summarized in a more broadly targeted book entitled *Margins of Reality: The Role of Consciousness in the Physical World*,[1] first published in 1987. Judging from the blizzard of personal responses from numerous readers who endeavored to share with us their own inexplicable experiences and perspectives, this popular representation also served the valuable role of reassuring these people of the validity of these events in the face of the rejection and ridicule to which they had been subjected by mainstream society.

Now, more than two decades later, the PEAR laboratory has concluded its university-based research operations and we feel obliged to complete the description of this nearly thirty-year research program. To this end, we offer this sequel volume in the hope that it will prove similarly beneficial in extending the base of scientific data and conceptual frameworks, further raising public awareness, and reassuring a latter-day complex of such sensitive readers. An assortment of experiments in human/machine interaction that have demonstrated the ability of volunteers to affect the performance of various random physical systems in accordance with their pre-stated intentions are summarized, along with the results of remote perception studies that have provided evidence that individuals are capable of acquiring information about geographical locations remote in distance and time without resort to the usual sensory inputs. These are followed by an array of theoretical models that attempt to illuminate the anomalous results and to propose a more comprehensive and powerful scientific and cultural paradigm that is capable of accommodating subjective factors in the establishment and representation of reality. Finally, we sketch the interpretations, principles, and conceptual vision that have emerged from this research and point to a number of pragmatic applications and that suggest new directions for scientific study.

While the initial challenge of fulfilling this constellation of tasks was somewhat overwhelming, the availability of detailed descriptions of our many experiments and theoretical efforts in published anthologies[2,3] and in the many articles and reports available on the PEAR website[4] have made it much more tractable. In addition, a useful supplement to these archival resources exists in the form of a tutorial DVD/CD set that captures the spirit and substance of the entire enterprise.[5] With all these supplementary accounts readily at hand, the tasks of preparing a concise and comprehensive update of our reservoir of empirical data and of effectively displaying the resultant expansions of our theoretical models and intellectual interpretations become somewhat less daunting, allowing us to build upward and outward by invoking generous references and extensions of these preceding representations.

Nonetheless, given the breadth of readership this work aspires to address, a significant issue remains: how much and what depth of tutorial technical material should be included. Here we have chosen to refer most of the technical matters to the referenced archival literature, distilling therefrom only the minimal details of process and nomenclature needed to make the surrounding prose comprehensible. We hope our readers will treat this format as more of a sampler or smorgasbord of the plethora of scientific, philosophical, and cultural issues raised by the research, and select those items that seem most relevant and stimulating to their particular backgrounds and interests.

The title of this new volume is intended to convey an important nuance in our research perspectives and aspirations as they have evolved. Namely, we now believe that the sundry anomalous physical phenomena that originally attracted our attention are deeply rooted in, and therefore significantly indicative of, a much more fundamental, profound, and ubiquitous metaphysical dynamic whose ultimate comprehension holds far richer potential for human benefit than the more explicit phenomenological curiosities with which we began. In fact, this deeper perspective is arguably more portentous than that of the prevailing scientific paradigm which it challenges. In allowing epistemological penetration beyond the superficial "margins" of reality into the depths of its essential "Source," the research and interpretations reported in this volume have provided a glimmer of a vast, poorly charted domain for future human exploration, comprehension, and utilization

For this text, we have eschewed such terms as "psychokinesis" or "PK" to describe the human/machine anomalies under study, in favor of language that more aptly conveys the *informational* character of the phenomena, *e.g.* "consciousness-related anomalies" or "consciousness-correlated physical phenomena." In fact, much of the traditional nomenclature of psychic phenomena has been set aside because of the increasingly antiquated conceptualization it implies.

More substantively, however, several additional topical perspectives have been added to the ensemble of contextual "vectors" that were originally proposed, such as the relevance of the work to the contemporary fields of biology and medicine, and its intrinsic roles in human creativity and spirituality. Likewise, the original technology theme has been appropriately expanded. Notwithstanding these updates, the philosophy, motivations, potential implications and applications that inspired inception of the PEAR program and have guided it throughout its odyssey have remained largely relevant.

Perhaps the most important ingredient to be emphasized in this new text, however, is the emerging promise of pragmatic applications of the principles, technology, and conceptual vision that have emerged from, and have been enabled by, the extant basic research accomplishments here and elsewhere. Regardless of how much inspiring, erudite, or credible scholarly reportage has appeared or may continue to emerge from such academic study, its ultimate impact will be far less than that of clear demonstrations of the practical utility of the phenomena. In the words of Norman Cousins, "The big ideas in this world cannot survive unless they come to life in the individual citizen."

References

[1] Robert G. Jahn and Brenda J. Dunne. *Margins of Reality: The Role of Consciousness in the Physical World.* San Diego, New York, London: Harcourt Brace Jovanovich, 1987. Reprinted: Princeton, NJ: ICRL Press, 2009.

[2] Robert G. Jahn and Brenda J. Dunne. *Two Decades of PEAR: An Anthology of Selected Publications.* Princeton, NJ: Princeton Engineering Anomalies Research, Princeton University, School of Engineering/ Applied Science, 1999.

[3] *EXPLORE: The Journal of Science and Healing, 3,* No. 3 (May/June 2007).

[4] <http://www.princeton.edu/~pear/>

[5] *The PEAR Proposition.* DVD/CD set produced by StripMindMedia and available on-line from the Publications page at ICRL <www.icrl.org>, via the appropriate link on the PEAR website, or at <http://www.amazon.com>.

SECTION I
Venues, Vistas, and Vectors

Vectors

1

CONTEMPLATING CONTEXT

All roads lead to Rome.
— Anonymous

The contemplation of consciousness, by its own definition, is a self-reflective process that is virtually impossible to approach objectively. As the ordering agent in the establishment of reality, consciousness expresses itself through our subjective experiences of the world around us, as perceived through an array of physical, psychological, and cultural filters that have been configured by our intentions, heritages, relationships, expectations, and previous experiences.[1] When we were preparing our first book, we therefore attempted to frame the results and inferences of the PEAR research program by first assembling a gridwork of relevant epistemological filters, or "vectors," that purported to sketch the long history and heritage of attention to consciousness-correlated anomalous phenomena akin to those on which our PEAR research was focused, and which offered an array of perspectives from which those findings might be considered. These elements included a mixture of philosophical, historical, and technical aspects that comprised a web of logic and heritage to justify and inform the experimental and theoretical strategies we were deploying. In brief review, these topics included:

Scientific Two-Step—an invocation of the long-revered dialogue of sound empirical experimentation with astute theoretical modeling, first proposed in the 17th century by Sir Francis Bacon as the "scientific method," with particular attention to the role of anomalies in motivating scientific investigations, and maintaining them on realistic and productive courses;

Man and the Mystical—an historical review of the human propensity to engage in a wide variety of spiritual, metaphysical, and ritualistic activities and beliefs, in efforts to understand and influence the establishment or perception of reality;

Scholarly Stream—a recitation of a few of the countless historical attempts to apply disciplined reasoning to the comprehension and utilization of metaphysical phenomena;

Parapsychological Perspective—a review and critique of this genre of professional attention to consciousness-related anomalous phenomena;

Critical Counterpoint—commentary on the role and value of critical assessment of scholarly work, and its abuses and failures;

Quantum Clues—the relevance of the perspectives, strategies, and nomenclature of quantum science in the study of consciousness-correlated physical phenomena;

Modern Man/Modern Machine—a prospectus on the utilization of modern technological resources in the research on consciousness-related anomalies, and the implications and possible applications of the results and understanding acquired therefrom in future engineering enterprises;

Statistical Science—elementary statistical concepts and formulas for evaluation of experimental results.

Each of these vectors reflected the prevailing state of our understanding, based on the empirical and theoretical conclusions we had reached at that point in our studies. Given the availability of *Margins* as a companion reference, little would be gained by further reprise of the details of these original vectors here. Much has changed in the subsequent decades, however, both in the monumental developments that have taken place in many fields of science and technology, and in the enhancement of knowledge acquired through continued laboratory investigations. While the array of the original perspectives continues to provide a relevant frame of reference within which to consider the implications of the PEAR research, a few extensions of that grid in those areas that bear on our thesis and have witnessed substantial expansion in recent years may prove useful additions to it.

Vectorial gridwork

In particular, four additional vectors became progressively more important to the study of consciousness over the past two decades: biology, medicine, creativity, and spirituality. But most notably, our original assessment of the status and projections of information technology needs to be expanded to encompass the many burgeoning arenas of engineering science that have emerged recently. Thus, the following chapters will add these emendations to the original list, before we attempt to summarize the results and implications of all three decades of our research.

Reference

---

[1] Robert G. Jahn and Brenda J. Dunne. "Sensors, filters, and the Source of reality." *Journal of Scientific Exploration, 18*, No. 4 (2004). pp. 547–570.

2

BIOLOGY AND EVOLUTION

Mind, rather than emerging as a late outgrowth
in the evolution of life, has existed always ...,
the source and condition of physical reality.
—George Wald[1]

One of the most longstanding dilemmas in the history and philosophy of science is the so-called mind/body paradox, which Arthur Schopenhauer referred to as "the world knot."[2] The problem resides in the difficulty of establishing a scientific connection between mental and physical phenomena that can specify in any useful detail how these two complementary categories of human experience relate to one another. Contemporary Western science tends to circumscribe its domain of intellectual authority in terms of objective physical criteria, leaving little room for subjective experiences that cannot be traced to specific physiological causes. In maintaining this position, it tends to overlook the self-evident fact that the primary work of science is the acquisition of empirical information and its representation in theoretical models. Yet both information acquisition and its representation are inherently mental processes whose exclusion from similar scientific inquiry severely penalizes the uniquely human enterprise of science itself. Mind without matter leaves us with a world of ephemeral abstraction; matter without mind eliminates the essence of life itself.

In the biological sciences, it is precisely here that we encounter the crux of the current intense debate between Darwinian evolution and Creationism. Despite the immense sophistication of contemporary research, development, and pragmatic deployment of the biological sciences that have devolved over their long, publicly pertinent histories, most recently exemplified in such subdivisions as biophysics, biochemistry, bioengineering, cellular and molecular biology, population biology, genetics, etc., the mechanism and taxonomy of the evolution of living species continues to divide this intellectual establishment. At this stage in their professional maturation an intellectual majority favors a predominantly secular option, variously labeled "neo-Darwinism," "natural selection," or "random mutation," wherein physiological improvement of a species is presumed to be driven by random adaptations to environmental conditions that utilize specific portions of its naturally occurring distributions of physical characteristics which, when propagated over many generations, preferentially aid in survival or dominance. But in an extreme opposing view, a vigorously committed cadre favors a more metaphysical model, variously labeled "Creationism," "Intelligent Design," or "Divine Intervention," which prefers to attribute the proliferate panorama of living creatures and their remarkably adept specialized capacities to the will and oversight of a supernatural deity of broadly varying cultural specifications.

An alternative hypothesis was proposed by Jean-Baptiste Lamarck, some 200 years ago, which has long been rejected as heresy by evolutionary theorists.[3] Lamarck, who died thirty years before Charles Darwin authored his *Origin of Species*,[4] claimed that acquired characteristics also could be inherited by offspring. But in addition he maintained that organisms have a tendency to evolve toward increasing complexity, through proactive responses to changing environments that require or encourage behavioral adaptation, not merely as the result of passive statistical reactions to random events. He referred to this process as *le pouvoir de la vie* (the life force) or *la force qui tend sans cesse à composer l'organisation* (the force that perpetually tends to make order). His hypotheses, however, have tended to be dismissed as scientifically inappropriate since they lack a basis in physical causality and are inconsistent with the entropy increase implicit in the Second Law of Thermodynamics.

More recently, however, a renewed interest has arisen in what is referred to as neo-Larmarckianism, which offers a middle ground between Darwinism and Creationism on which these polar factions might rationally meet. In this approach, it is argued that if the species in question were to embody some form of independently creative consciousness, and if that consciousness were indeed allowed some teleological capacity, *i.e.* if a species' evolution were to be driven, at least to some extent, by its own desires or self-evident needs, this would, *de facto*, constitute a broadly distributed "intelligent design" mechanism that still utilized the random distribution of characteristics of the species to accomplish its evolutionary goals, while at the same time constituting what could be regarded as a broadly distributed form of "Divine Plan."

Encouraged by a growing body of empirical evidence emerging from the modern fields of genetics and molecular biology,[4-7] Italian biologist Antonio Giuditta has proposed a "spiral mechanism of evolution" that involves a bi-directional flow of information between the environment and the organism that can affect the genome of an evolving species.[8] In several of his writings he raises the profound possibility that "the remarkable computing capacities of the organism's integrative functions becomes a plausible explanatory option" for the coordination of its response to the environment, which utilizes autonomic complex algorithms that modulate and control performance at all levels of organization, from molecules to behavior.[9] Elsewhere, Giuditta observes that "the two major tenets of neo-Darwinism, random mutation and natural selection, cannot readily explain the birth of complexity and even less its progressive phylogenetic increment and its associated progressive increase in the

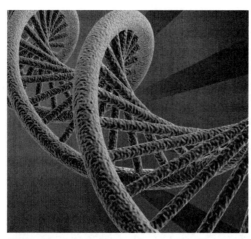

DNA spiral

autonomy of organisms."[10] According to this model, the properties of the constituent units of an organism are modulated in part by their inclusion in a more complex system whose collective features are not evident in the component units.

A major obstacle to full development of this theory is the issue of how this internal information can be transmitted and shared at all levels of biological organization, from the most elementary components to the complex whole. One such mechanism, proposed by biophysicist Fritz-A. Popp and others,[11-13] suggests that this may be accomplished through the exchange of light, in the form of biophotons: optical quanta that are continuously emitted from and absorbed by all biological systems and sub-systems that are far from thermal equilibrium. Although this radiation is of very low intensity, it can be detected and measured, and it displays some unexpected and important properties. Among these are its tendency to be emitted in a coherent, laser-like fashion, whereby the sub-systems act collectively to signal the state of the entire emitting organism, much like an ongoing broadcast of its current status. Minute changes in the physiological and environmental conditions of the system are reflected in its biophoton signal, which even can give indication of its biological quantum state. This characteristic coherence obtains only in the emissions of living systems, and the degree of coherence is an indicator of the health of the organism. For example, the eggs of free-range chickens have been shown to display higher intensities of biophoton emission than those of their caged factory relatives, and cancer cells have been found to emit a less coherent spectral distribution of photons indicative of increasing entropy, such as one would find in a non-living closed system.[14]

According to Popp, biophotons play a central role in the basic regulation of biological functions, cell growth and differentiation, connections to delayed luminescence, other forms of biological coherence, and supermolecular processes in living tissues. In his essay entitled "The Power of Weakness,"[15] Popp notes:

Over the years it has been possible to demonstrate that bio-photons are signals of intercellular communication, with the coherence of the emissions serving as indicators of this information exchange. While technically non-functional, at a sufficiently high level coherence is capable of enhancing a system's performance; producing awareness of long-distance interactions (which play an important role in nonlinear relationships through the overlapping of information and randomness); and in developing what we may call "consciousness" in internal structures. In other words, consciousness itself can be regarded as an evolutionary process based on coherent states, raising a completely new perspective of how information is filtered from the ocean of potential information and how coherence among individual consciousnesses can affect the perception and integration of that shared information.

His view on the origin and evolution of mind is consistent with Giuditta's observation in his "Essay on the Nature of Mind":

By the progressive differentiation of elementary particles into molecules, of primitive cells into multicellular organisms, all the way up to the astonishing structure of brain, it is conceivable that energy fields may have evolved from elementary mental entities to attain the qualities and capacities of human mind.[16]

Such biophotons also may pose a potential mechanism for Rupert Sheldrake's hypothesis of "morphic fields" that underlie our perceptions and mental activities and impose patterns on otherwise random or indeterminate processes.[17] Sheldrake further suggests that these morphic fields also function in social groups, providing non-local channels of communication and connecting members of a group even when they are physically separated.

The possibility of a teleological capacity of consciousness, as proposed in the neo-Lamarckian model, predicates a more generic and profound assessment of the nature and role of consciousness *per se* in the life process, and the relevance of this to our PEAR research has become progressively more evident and fundamental. It is our conviction that scholarly exploration of this terrain will require the intellectual traveler to stride past the prevailing physicalistic view of the mind as an epiphenomenon of the brain structure and its chemical, physical, and neuronal communication networks. Without dismissing lightly the immense advances in our understanding of brain function that have emerged from CAT scans, PET scans, fMRIs, and related technologies, we must give advance warning that the research reported in the following sections of this book, along with insights derived from many other sources in many other venues, simply cannot support such a mind/brain isomorphism, and we shall need to struggle on into yet more complex intellectual territory.

It is a territory that complicates the prevailing epistemological architecture of objective, deterministic, space/time causation by acknowledging a pro-active role for consciousness, along with a host of subjective experiential correlates that are capable of functioning independently of conventional spatial and temporal constraints. Even more striking, it is this territory that allows some form of "consciousness" to be ascribed to all living creatures, however rudimentary, and even to objects and substances normally regarded as devoid of life but which are capable of exchanging information with their environments. Hardly a familiar or comfortable terrain, but our empirical results cannot help but lead us deeply into it, where we shall find that our biological vector requires an ambipolar generalization, *i.e.*, we now need to confront symmetrical twin mysteries: "The role of biology in the world of consciousness;" and "The role of consciousness in the biological world." This was beautifully illustrated in an anecdote from a lecture given by the late Nobel laureate and renowned environmental biologist, George Wald:

Some years ago a French biologist named Fauret-Frémiet made a fine film on the feeding behavior of herbivorous and carnivorous ciliate protozoa. Such protozoa, those single-celled animals, go about, digest, reproduce, and they do everything a multi-cellular animal does. Watching Fauret-Frémiet's movie, I could not keep myself from anthropomorphizing. These protozoa were frequently doing exactly what you would do in the same predicament. I remember particularly a protozoan going at a microscopic piece of meat. It fastened onto one of the muscle fibers and then backed up to pull it away, and it would not come. So it went in again, and backed up at another angle, and went in again and backed up at another angle, worrying off this fiber. And I was thinking, that is exactly what a dog would do, trying to deal with a chunk of meat. Was this single-cell aware of its activity and the purpose of that activity? For that matter, is a dog aware? Again, no scientific confirmation or disconfirmation of the question is possible.[18]

As we pursue our own journey into this exotic territory, we must keep our highly developed analytical minds open to instructive empirical indications of far less familiar biological functions that may at the end of the day provide the missing links for comprehension of the consciousness-related anomalous capabilities we seek to understand.

References

[1] George Wald. As quoted in *Bulletin of the Foundation for Mind-Being Research*. Los Altos, CA; *3* (September 1988).

[2] Arthur Schopenhauer. *Pererga and Paralipomena*. Trans. Thomas Bailey Saunders. Hayn: Berlin, 1851. Trans Eric F. J. Payne. Oxford: Clarendon Press, 1974.

[3] Jean-Baptiste Pierre Antoine de Monet de Lamarck. *Philosophie zoologique, ou, Exposition des considérations relatives à l'histoire naturelle des animaux*. Paris: Chez Dentu [et] L'Auteur, 1809.

[4] Charles Robert Darwin. *The Origin of Species by Means of Natural Selection*. London: John Murray, 1859.

[5] L. Brent, Lee S. Rayfield, P. Chandler, W. Fierz, P. B. Medawar, and Z. Simpson. "Supposed Lamarckian inheritance of immunological tolerance." *Nature*, *290*, No. 5806 (1981). pp. 508–512.

[6] Alan Durrant. "The environmental induction of heritable change in *Linum*." *Heredity*, *17*, No. 1 (1962). pp. 27–62.

[7] B. C. Goodwin and G. C. Webster. "Rethinking the origin of species by natural selection." *Rivista di Biologia*, *74* (1981), II-26. [Riv.Biol. 1999 Sep–Dec;92(3):464–7.]

[8] Edward J. Steele. *Somatic Selection and Adaptive Evolution: On the Inheritance of Acquired Characters*. Toronto: Williams & Wallace, 1979.

[9] Antonio Giuditta. "Proposal of a 'spiral' mechanism of evolution." *Rivista di Biologia*, *75* (1982). pp. 13–31.

[10] Antonio Giuditta. "Mind and Biological Evolution." In Zachary Jones, Brenda Dunne, Elissa Hoeger, and Robert Jahn, eds., *Filters and Reflections: Perspectives on Reality*. Princeton, NJ: ICRL Press, 2009. pp. 177–186.

[11] Antonio Giuditta. "Creative Evolution: What Are Your Mechanisms?" Proceedings of the 7th European Meeting of the Society for Scientific Exploration, Röros, Norway, August 17–19, 2007. pp. 147–150.

[12] Rajendra Prasad Bajpai. "Coherent Nature of Biophotons: Experimental Evidence and Phenomenological Model." In Jii-Ju Chang, Joachim Fisch, and Fritz-Albert Popp, eds., *Biophotons*. Dordrecht: Kluwer Academic Publishers, 1998. pp. 323–339.

[13] Fritz-Albert Popp. "Evolution as the Expansion of Coherent States." In Fritz-Albert Popp, Kathleen Hung Li, and Qiao Gu, eds., *Recent Advances in Biophoton Research and Its Applications*. Singapore, New Jersey, London, and Hong Kong: World Scientific, 1992. pp. 445–456.

[14] Fritz-Albert Popp. "Quantum Phenomena of Biological Systems as Documented by Biophotonics." In Avshalom C. Elitzur, Shahar Dolev, and Nancy Kolenda, eds., *Quo Vadis Quantum Mechanics?* New York: Springer, The Frontiers Collection, 2005. pp. 371–396.

[15] Klaus Lambing. "Biophoton Measurement as a Supplement to the Conventional Consideration of Food Quality." In Fritz-Albert Popp, Kathleen Hung Li, and Qiao Gu, eds., *Recent Advances in Biophoton Research and Its Applications*. Singapore, New Jersey, London, and Hong Kong: World Scientific, 1992. pp. 393–413.

[16] Fritz-Albert Popp. "The Power of Weakness." Foreword to Zachary Jones, Brenda Dunne, Elissa Hoeger, and Robert Jahn, eds., *Filters and Reflections: Perspectives on Reality*. Princeton, NJ: ICRL Press, 2009. pp. xiii–xvi.

[17] Antonio Giuditta. "Essay on the nature of mind." *Rivista di Biologia*, 97, No. 2 (2004). pp. 187–196.

[18] Rupert Sheldrake. *A New Science of Life: The Hypothesis of Formative Causation*. Los Angeles, CA: J. P. Tarcher, 1981.

[19] George Wald. "Consciousness and Cosmology." In Andrew C. Papanicolaou and Pete A. Y. Gunter, eds., *Bergson and Modern Thought: Towards a Unified Science*. New York: Harwood Academic Publishers, 1987. p. 349.

3

MEDICAL MYSTERIES

In medicine, following the lead of physics, we have striven to exclude subjectivity wherever it has arisen, which has meant denying a role for the mind and meaning in health and illness.
— Larry Dossey[1]

Notwithstanding all of the sophistication of modern medical practice and the evident benefits of its diagnostic technologies, pharmaceutical remedies, and elaborate surgical capabilities, these have tended to become increasingly impersonal and encumbered by the less-than-comprehensive, less-than-compassionate, protocols and techniques it has been driven to adopt by the cultural, financial, legal, and political pressures that bear upon it. Few of us have been spared intense, often poignant, involvements with the contemporary medical establishment, some of us to benefit from the wealth of experience, resources, and facilities it can provide, others to suffer from the generic approaches that inevitably attend its monumental bureaucratic architectures.

Disillusionment with the perceived inaccessibility, impersonality, cost, and in many cases, ineffectiveness of the contemporary allopathic approaches has driven growing communities of consumers to explore and utilize a burgeoning array of alternative health care options, many of which entail reinvestment in ancient or remote cultural traditions or follow more intuitive or humanistic practices than their establishment counterparts. These have been subsumed under categories of "complementary," "integrative," or simply "alternative" medicine, and include such practices as homeopathy, acupuncture, yoga, meditation, herbal remedies, prayer, and contemporary shamanism, among others, that until quite recently have been largely dismissed by mainstream medicine as scientifically untenable folklore, despite their demonstrated effectiveness in many cases.

Yet, mainstream medicine itself hides in its stockroom of empirical experience an array of anomalies that blatantly defy causal connections, diagnoses, and treatments, and which surely signal the existence of enduring fundamental mistakes or inadequacies in its prevailing reductionist paradigm. Consider just this short list:

Placebo and nocebo effects—"Sugar pills" or therapeutic techniques that with no identifiable pharmacological, neurological, or functional basis, seem to relieve or totally cure a variety of physiological complaints or, in other cases, to stimulate or exacerbate them;

Psychoneuroimmunology—Palpable revisions of allergic or immune responses associated with psychological factors;

Multiple Personality Syndromes—Spontaneously occurring or psychologically induced major alterations of personality profiles that can be accompanied by comparably large shifts in various physiological characteristics;

"Black Monday" Syndrome—Evidence that more fatal heart attacks and strokes occur around 9:00 am on Mondays than on any other day of the week, indicating a stronger correlation with job dissatisfaction than with any other known physical cause;[2]

Spontaneous Remissions—Unanticipated and implacable disappearances of serious medical symptoms, most notably many forms of cancer, not attributable to allopathic treatments;

Prayer Therapy and Energy Healing—Demonstrable improvements in healing rates and outcomes in patients addressed by individual or group prayer, the laying on of hands, or the non-local transmission of some form of "life energy," compared to control samples of unaddressed patients suffering similar maladies.

Other such medical enigmas could be cited, several of which have been targeted for study by credible organizations, investigators, authors, and private individuals, but have yielded little useful rationalization within the standard medical lexicon and its prevailing repertoires of diagnosis and treatment. And there are numerous examples from other cultures of effective shamanic and alternative healing traditions that are completely at odds with the allopathic model that dominates the Western medical paradigm. The only common denominator that seems to characterize this disparate assortment is some proactive involvement of the human psyche in the formation of physiological response to subtle and explicit forms of *information,* an ingredient not systematically included in the customary healing recipes. The efficacy of homeopathy, placebos, and psychotherapeutic and shamanic techniques that derive from the emotional resonance between therapist and client; the various strategies of so-called "energy healing;" the non-local influence of prayer; all attest to the role of information in re-establishing the patient's natural capacity for self-regulation, and thereby to the active participation of consciousness in the recovery and maintenance of physical health. Conversely, the negative impact of depressing diagnostic information on the physiological course of disease or immune response is widely experienced and recognized. We seem to have forgotten that the mind and body, *psyche* and *soma,* are inextricably intertwined, and that which affects one aspect of our being will inevitably affect the other.

There is a certain irony in the medical profession's continuing display as its official symbol the caduceus of Hermes, Greek god of commerce; guide of the dead; protector of merchants, gamblers, liars, and thieves; and inventor of magical incantations; with its two serpents entwined in the form of a double

Caduceus

helix. It seems far less apt than the more traditional Greek symbol of healing, the rod of Asclepius with its single, wingless snake.

*Asclepius'
staff*

While it is far from our purpose to survey the prevailing panoply of conventional and alternative medical practices, let alone to attempt to rank their relative effectiveness, it will be evident from the following text that much of the experimental and theoretical research on consciousness-related anomalous behavior of physical devices and systems also bears considerable relevance to the anomalous effects that appear within the context of contemporary and future health care. In particular, such subjective factors as intention, resonance, and attitude, which appear to be crucial in the manifestation of our engineering anomalies, are the very features that frequently are lacking in modern medical practice but characterize many of the alternative treatments that prove to be effective healing agents. It was this recognition that prompted invitation and publication of a special issue of *EXPLORE: The Journal of Science and Healing*, which presents an anthology of PEAR publications relevant to the subject of health,[3] and of other articles and reports dealing with this relationship.[4-8]

We once presented a talk at a conference on "Spirituality and Health Care" to a large audience of physicians, nurses, and alternative health care providers. At that time, we were unsure how our basic research on human/machine anomalies and remote perception would relate to this theme or to this forum, so we were somewhat surprised when its presentation prompted an extended standing ovation. At the close of the session, we queried the organizer why this audience seemed so enthusiastic about a lecture on laboratory studies of consciousness-related anomalies. He responded that medical practitioners encountered all sorts of inexplicable physical events in the course of their work, but lacking any scientific basis for their confirmation, such experiences are not officially credited; rather, they are typically discounted as anecdotes. Thus, our findings had provided some empirical verification of this dimension of effective medical treatment. In particular, the audience seemed to have resonated with a quotation from *Margins of Reality*:

> ... successful strategy for anomalies experimentation involves some blurring of identities between operator and machine, or between percipient and agent. And, of course, this is also the recipe for any form of love: the surrender of self-centered interests of the partners in favor of the pair.[9]

This, it appears, is consonant with the intuitive techniques of many successful healers, who find themselves in some cognitive dissonance with respect to their formal medical training. It is the generic form of this escalating cultural disparity between analytical precision and intuitive resonance, between head and heart, between logic and love, that the PEAR program has attempted to address.

References

[1] Larry Dossey. *Meaning and Medicine: A Doctor's Tales of Breakthrough and Healing*. New York, Toronto, London, Sydney, Auckland: Bantam Books, 1991. p. 12.

[2] Gina Kolata. "Heart attacks at 9:00 A.M." *Science* 233 (July 25, 1986). pp. 417–418.

[3] *EXPLORE: The Journal of Science and Healing*, *3*, No. 3 (May/June 2007).

[4] Robert G. Jahn. "Information, consciousness, and health." *Alternative Therapies*, *2*, No. 3 (1996). pp. 32–38.

[5] Brenda J. Dunne and Robert G. Jahn. "Consciousness, information, and living systems." *Cellular & Molecular Biology, 51,* No. 7 (2005). pp. 703–714.

[6] Robert G. Jahn. "'Out of this Aboriginal Sensible Muchness': Consciousness, information, and human health." *Journal of the American Society for Psychical Research, 89,* No. 4 (Oct. 1995). pp. 301–312.

[7] Roger D. Nelson. "The Physical Basis of Intentional Healing Systems." Technical Note 99001. Princeton Engineering Anomalies Research, Princeton University, School of Engineering/Applied Science, Princeton, NJ. January 1999 (28 pages).

[8] Robert G. Jahn, Brenda J. Dunne, and York H. Dobyns. "Exploring the Possible Effects of Johrei Techniques on the Behavior of Random Physical Systems." Technical Report PEAR 2006.01. Princeton Engineering Anomalies Research, Princeton University, School of Engineering/Applied Science, Princeton, NJ. January 2006 (30 pages).

[9] Robert G. Jahn and Brenda J. Dunne. *Margins of Reality: The Role of Consciousness in the Physical World.* San Diego, New York, London: Harcourt Brace Jovanovich, 1987. Reprinted: Princeton, NJ: ICRL Press, 2009. p. 343.

4

CONSCIOUSNESS AND CREATIVITY

Every child is an artist. The problem is how
to remain an artist once he grows up.
— Pablo Picasso

Creativity is consciousness at play—teasing apart preconceptions, recombining ideas and images in novel ways, and spinning metaphors to represent inner experiences in the tangible world. One might also define creativity as the ability to bring something new into existence or to invest something that already exists with a new form, through the play of imagination. In this view it is not surprising to encounter the concept of play in many creative contexts: we play a musical instrument, we act in a play, we play with words, and we recreate via play.

Rather than utilizing the deductive logic required by analytical activities, creativity draws its primary inspiration from new associations among existing concepts and images to generate new and innovative ways of perceiving and representing experience. In contrast to the more linear logic of the rational mind, the creative consciousness tends to deal in symbols and metaphors, sounds and colors, shapes and movements. It speaks to us at the intuitive and emotional level of our being, touches on the mystical, embraces the irrational, and thrives on uncertainty. Creativity can express itself in all dimensions of human experience and readily tolerates and utilizes acausal, non-linear, and intuitive processing to complement its logical capacities. In this way it can avail itself of the random fluctuations inherent in the physical world, even though its mystical heritage and its sometimes imprecise forms of expression make the creative process difficult to describe or quantify, and thus less amenable to rigorous scientific analysis.

Art is one form of creativity that includes both abstraction and specification, the interplay of which is evident in the work of many of the greatest thinkers and artists over the course of human history. The scientific accomplishments of such luminaries as da Vinci, von Goethe, or Swedenborg were comparable with their monumental artistic and literary achievements. Notwithstanding, there has been a tendency on the part of some scientific scholars to regard art somewhat pejoratively, as the antithesis of science rather than as its partner in the exploration and understanding of nature. Others have better appreciated this relationship, Albert Einstein for example, who noted that:

> All religions, arts and sciences are branches of the same tree. All these aspirations are directed toward ennobling man's life, lifting it from the sphere of mere physical existence and leading the individual towards freedom.[1]

Attempts to separate these intrinsically interwoven aspects of human experience are ultimately unproductive. Although we have come to speak in terms of a "right brain" and a "left brain," in fact we have but one brain with complementary capabilities, both of which are essential for a fully functioning consciousness. Denying one in favor of the other demeans both. This point was beautifully represented in Richard Strauss's last opera, "Capriccio," wherein he posed the irresolvable question of the primacy of words versus music in operatic composition.

Art and science have evolved together, driven by the same creative impetus, both of them reflecting prevailing cultural trends and technological advances. The ancient Greek term *technê* (art or craft) stems from the same root as the word *technology*, and until the late 18th century art was a term used to describe both highly skilled and sublime practices. Thus technology, the advanced use of tools, and the creation of things like houses, sculptures, poems, or music, were conceptually connected throughout most of human history. Science and art both are creative approaches to understanding the world: the former follows an objective path of empirical and analytical investigation that breaks things down to their constituent parts; the latter follows a subjective path of synthesizing observations and feelings to create new meanings through symbolic and aesthetic representations. These two strategies are complementary, and both are essential for any act of creativity or any comprehensive understanding of reality.

Historically, the creative human spirit has provided innovative developments in both art and science that have reflected one another and, in turn, stimulated social and cultural change. For example, the imaginative technological developments that drove the Industrial Revolution stimulated new forms of expression in art, music, and literature that tracked the evolving social ethics and aspirations of that era. Modern computer technology now permits us to enjoy classical works of art that previously were accessible only in museums, libraries, and concert halls, and computer-generated art and music have added vast new dimensions to these disciplines. As we become aware of more abstract properties of our world, we need more refined methods to represent those concepts. As always, we experience the world through our senses, but now we tend to evaluate and analyze those experiences using mathematics and other forms of abstraction. But in order to savor these abstractions, we need to experience them once again through our senses, and for that we need to return to our art forms. While science provides intellectual representations of our knowledge, art demonstrates these complicated representations in terms that speak to the soul. We only truly understand the world when we can feel it as well as specify it.

One example of such creative blending of art, science, and technology found expression in an innovative collaboration with the Danish artists, Christian Skeel and Morten Skriver, which resulted in an installation displayed at the Trapholt Museum in Denmark. As an amalgamation of their artistic vision with our random event generator (REG) technology, the project was designed to engage the visitor on a personally meaningful level and, at the same time, provide empirical data on human/machine interactions. Viewers were presented with an evolving image of a baby, which ranged from an indistinguishable pattern of random pixels through various degrees of clarity, based on the output of the REG.

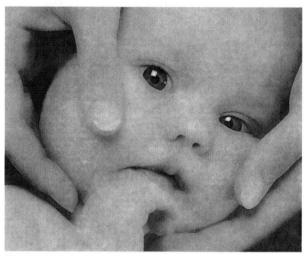

Trapholt experiment image

From a scientific perspective, the Trapholt Experiment was an investigation of the effects of consciousness on a microelectronic random process; as a work of art, it drew the viewer into the archetypal, magical roots of visual art. It thus linked the museum to the laboratory and to the temple. By definition, the result of the experiment was uncertain, and allowed many interpretations, and it was precisely in this uncertainty and range of interpretation that art and science met. The installation was received with strong public acclaim and positive reviews from art critics. It also stimulated the interest of physicists from the prestigious Niels Bohr Institute in Copenhagen, who undertook to develop a second-generation rendering of the installation, and to engage with the artists in another art exhibit that displayed the patterns created by wave interference in the sand at the edge of a beach.

The combination of high-precision technical equipment and the patterns of drifting sand produced an engagingly poetic statement about the meeting between the analytical scientific eye and the mysteries of nature. A similar example of random event generator–based art was exhibited at The Hague in 2008 by Dutch artist Iebele Abel.

In describing their vision of such integration of art and science, Skeel and Skriver articulated the importance of a balanced relationship between these two aspects of our human being:

> Neither art nor science can exist alone. They need each other the way the spirit needs matter in order to create a universe that can be seen, heard, and tasted. Art and science represent two complementary sides of man. So they also possess fundamental similarities. In a sense their goals are identical. They both employ intuition and intelligence to get closer to the mystery of existence. They try to create images that make sense. So the two areas must learn to talk and work together.[2]

In Abel's terms:

> Art originates through perception and thinking about perception. Art does not differ substantially from science in this sense. In both disciplines perception and thinking pose new questions again and again; questions that, so it seems, never let themselves be answered completely.[3]

While visual artistic representation is perhaps the most explicit form of creative expression, it is by no means the only one. Beyond it and its literary, musical, and dramatic siblings, creativity vitalizes in all manner of scientific enterprises, business endeavors, political stratagems, and individual personal affairs. The most enlightening scientific experiments and theoretical models are distinguished from the more pedestrian versions by the sweeps of intellectual creativity they capture. The most profitable and popular business enterprises require, recognize, and solicit unceasing streams of creative initiatives in their managements and implementations. Effective leadership depends on the persuasiveness with which individuals and organizations can represent themselves and their agenda priorities. And the interpersonal attractiveness of individuals usually reflects their creative responses to challenges and opportunities. While it is widely conceded that a solid technical education is a good preparation for successful careers in management of virtually any productive enterprise, it is no less demonstrable that a broad liberal education that encourages creative thinking and intuitive acumen over a wide spectrum of venues frequently endows the most successful top managers in their respective fields.

It should become clear that the anomalous effects and experiences derived from our research, their comprehension, and their utilization, are closely coupled to, indeed enabled by, a very similar capacity for non-linear thinking, imagination, and subjective identification and involvement that characterize consciousness and creativity in their host of more conventional deployments. No better indication can be found, therefore, that the anomalous events themselves are normal epiphenomena of the creative human minds that experience them.

References

[1] Albert Einstein. *Out of My Later Years*. New York: Philosophical Library, 1950.

[2] Christian Skeel and Morten Skriver. "Art, Mind, and Matter." In Zachary Jones, Brenda Dunne, Elissa Hoeger, and Robert Jahn, eds., *Filters and Reflections: Perspectives on Reality*. Princeton: ICRL Press, 2009. pp. 75–90.

[3] Iebele Abel. *Talks about Mind over Matter.* The Hague: Elmtree and Waters, 2010. p. 7.

5

THE SPIRITUAL SUBSTANCE
OF SCIENCE

Even as the finite encloses an infinite series
And in the unlimited limits appear,
So the soul of immensity dwells in minutia.
And in narrowest limits no limit inhere.
What joy to discern the minute in infinity!
The vast to perceive in the small, what divinity!
— Jakob Bernoulli[1]

The human being is an inherently paradoxical creature, simultaneously drawn in two directions. On the one hand, it is an intrinsically spiritual entity that aspires to express itself in physical form; conversely, it is a complex biological organism yearning for spiritual transcendence. While this dynamic probably prevails to some degree in all living entities, most seem to integrate these propensities within an instinctive complementarity that typically presents little conflict. The human species, however, tends to render this interplay into a dualistic opposition, resulting in an endemic cognitive dissonance and in a predominantly physicalistic approach to life, wherein death is dreaded as a foreign and frightening menace. It may well be that this bifurcation of the natural fullness of being is what most uniquely characterizes our species, more so than our highly touted linguistic, tool-making, or cognitive abilities.

The foundations of Western society, most notably its science, actually derive from ancient metaphysical principles and, despite their contemporary secularity, remain rooted in a primordial desire to comprehend, conquer, and control the unknown. Over the larger portion of human history and across widely diverse cultures these existential features have been explicitly ascribed to various forms of divine order, and the quest for their understanding inevitably has entailed spiritual as well as technical dimensions. From early Greeks to medieval alchemists to Renaissance astronomers, the primary scientific task was to bring the mind of the scholar into resonance with the Source of being. The ancient Hermetic principle, "What is below is like that which is above"[2] led Plato to postulate "Know thyself" as the major precept of his philosophy and those of many to follow. What we have labeled the "Scientific Revolution" was instigated by scholars steeped in alchemical traditions who engaged in heated debates regarding subtle metaphysical issues and nuances. For example, in the early 17th century Johannes Kepler and Thomas Fludd argued extensively about the

Geocentric representation of the solar system

geometry of planetary orbits on the basis of the mystical meanings of numbers and the relationship between macrocosm and microcosm. While Kepler sought to derive a mathematical description of the cosmos, Fludd's view of the world encompassed the human mind as well as the physical world.[3]

Nearly a century later, despite their well-known controversy regarding the invention of the calculus, both Isaac Newton and Gottfried Leibniz acknowledged the essential relationship between mind and matter. Newton regarded space and time as a category of spirit, and the ultimate mechanism of change in the universe to reside in "the mystery by which mind could control matter;"[4] Leibniz asserted that "The soul follows its own laws, and the body likewise follows its own laws; and they agree with each other in virtue of the pre-established harmony between all substances,

since they are all representations of one and the same
universe."[5] Notwithstanding,
these scholars differed with
regard to whether this relationship could be demon-
strated by pure mathematical logic or whether it entailed an experiential component as well.

Since the 17th century,   *Alchemist at work*
however, Cartesian dualism,
although itself derived from metaphysical principles, has emphasized a distinction between mind and matter that has steadily
driven science to focus on matter far more than on mind, and on
analytical applications of its empirical principles far more than on
subjective experience. The ramifications of this spirit/substance
disparity cannot be overstated; acknowledged or not, they bear
on virtually every aspect of our contemporary society: science,
philosophy, education, politics, economics, healthcare, and even
religion, to the extent that our worldview has come to place a disproportionate emphasis on physical reductionism and causality,
and to trivialize the intangible and immeasurable aspects of experiential reality. But as in most natural processes, disequilibrium
tends toward self-correction, with or without conscious human
assistance. This compensation may express itself physically as disease or emotional disorder on an individual level, or as economic
uncertainty, social upheaval, or political insecurity on a societal
level. Even on the global scale, we are now seeing clear evidence
that a predominantly materialistic worldview can severely disrupt
the balances of nature itself.

During such periods of high uncertainty, when familiar patterns of thought and behavior seem to lose their effectiveness and
authority, the human tendency is to turn for reassurance back
to its spiritual roots. The individual trapped in a life that seems

to have lost its meaning frequently turns to prayer, meditation, or some other spiritual path to find solace or inner guidance; a troubled society displays trends toward intensified religious fundamentalism, a return to archaic metaphysical traditions, or the spawning of an assortment of new mystical practices.

Such periodically alternating cultural trends are frequently characterized by renewed interest in supernatural phenomena.[6] In the 17th century, for example, when the metaphysical and experiential models of Leibniz and Goethe were struggling for dominance with the more reductionistic and pragmatic approaches of Newton and Decartes, this surfaced in cultural preoccupations with Hermetic and alchemical traditions.[7] It re-emerged again during the Industrial Revolution, toward the end of the 19th century, when technology and materialism began to dominate western culture and a popular fascination with spiritualism arose, prompting leading scholars in Europe and the United States to explore mediumship and other forms of anomalous occurrences and to apply the methods of science in a search for evidence of post-mortem survival. (One of the most comprehensive records of these explorations is F. W. H. Myers' posthumously published classic, *Human Personality and Its Survival of Bodily Death*, which describes a vast array of phenomena of subliminal origin, including dreams, automatic writing, hallucinations, apparitions, creativity and genius, hysteria, multiple personality, trance mediumship, telepathy, hypnosis, and mesmerism.[8]) The renowned Harvard psychologist and philosopher, William James, recognized the critical role of such anomalous phenomena in his landmark book, *Varieties of Religious Experience*,[9] and his colleague and successor, William McDougall, was instrumental in establishing the first parapsychology laboratory at Duke University in the late 1920s.

It may not be coincidental that the arcane philosophical debates of the 17th century led to the intellectual innovations of the "Scientific Revolution," and that those of the 19th century were precursors to quantum science and relativity, along with the

Freudian concept of the "unconscious." In both eras, the new perspectives on the nature of reality led to the development of new technologies and resultant new empirical evidence that challenged the prevalent ways of thinking.

At the present time, we are again immersed in a period of economic uncertainty and political insecurity that is stimulating greater participation in fundamental religions and less conventional spiritual practices. The dispute between proponents of Darwinian evolution and "intelligent design" mentioned earlier is only one example of the increasing tension between physical and metaphysical representations of reality. Another is the growing awareness and passionate discussions regarding global warming and the broader role of human intervention in the processes of nature. A new generation of scientific tools has raised considerations of the informational nature of the physical world that have not previously been accessible, and science continues to probe increasingly abstract and counter-intuitive domains of probabilistic quantum mechanics, special and general relativity, string theory, and nonlinear dynamics. The role of spirit—whether divine or human, individual or collective—in the structure and operation of the physical world is now returning to theoretical and pragmatic relevance and, despite the resistance of those who remain committed to a materialistic and reductionistic representation of reality, is becoming increasingly difficult to dismiss. And along with all of this, we are once again witnessing a revival of popular interest in metaphysics, spirituality, and a host of anomalous phenomena.

This recurring cycle is metaphorically suggestive of an underground stream that flows unseen below the surface of our rational preoccupations with pragmatic and intellectual activities, until those begin to show signs of inconsistency, fragility, and undependability. At such times it burbles to the surface, revealing our spiritual insecurities, replete with their ambiguous, irrational, irreproducible anomalies, and contributing to our uncertainties with inexplicable, yet undeniable, evidence of a deeper source of meaning. It irrigates our ground of being, fertilizing the creative process and challenging us to reconsider our assumptions, revise our theories, and reassert our spiritual heritage and purpose.

The empirical evidence for the non-local nature of consciousness emerging from the research at PEAR and elsewhere inescapably predicates questions about the non-physical spiritual dimensions of human experience and offers a critical nexus in the potential for reuniting science, substance, and spirit. In our earlier book, we offered another metaphorical representation of the spirit/substance dynamical relationship in terms of an informational wave function, which we associated with "consciousness," that establishes standing waves of experience within an environment of a finite physical potential well. This view bypasses the prevailing "mind/body problem" by offering an alternative perspective that embraces both spirit and substance in a mutually creative relationship that engenders all human experience. The watershed issue facing us today is whether we have reached sufficient intellectual maturity to include within our contemporary worldview a model that can accommodate the essential dynamic interplay of the material and spiritual aspects of human nature that has always been implicit in our intuitive conceptualizations of reality.

References

[1] Jakob Bernoulli. *Ars conjectandi: Tractatus de Seriebus Infinitis,* 1689, p. 306. (Quoted in Frank J. Swetz, ed., *From Five Fingers to Infinity: A Journey through the History of Mathematics.* Chicago: Open Court, 1996.)

[2] Hermes Trismegistus. Emerald Tablet. In Julius Ruska, *Tabula Smaragdina: ein Beitrag zur Geschichte der hermetischen Literatur.* Heidelberg: Carl Winter's Universitätsbuchhandlung, 1926. p. 2.

[3] Wolfgang Pauli. "The Influence of Archetypal Ideas on the Scientific Theories of Kepler." In Carl Jung and Wolfgang Pauli, eds., Priscilla Silz, trans., *The Interpretation of Nature and the Psyche.* New York: Pantheon Books, 1955. Bollingen Series L1. pp. 121–131.

[4] David Kubrin. "Newton's Inside Out! Magic, Class Struggle, and the Rise of Mechanism in the West." In Harry Woolf, ed., *The Analytic Spirit: Essays in the History of Science in Honor of Henry Guerlac.* Ithaca and London: Cornell University Press, 1981. p. 113.

[5] Gottfried Wilhelm Leibniz. *Monadology,* #78, 1714. Translated by Robert Latta. (Republished 2008 by Forgotten Books.)

[6] Stephen G. Brush. *The Temperature of History: Phases of Science and Culture in the Nineteenth Century.* New York: Burt Franklin and Company, 1978.

[7] Frances Amelia Yates. *The Rosicrucian Enlightenment.* Boulder, CO: Shambhala, 1978.

[8] Frederic William Henry Myers. *Human Personality and Its Survival of Bodily Death.* New York: Longmans, Green, 1907.

[9] William James. *Varieties of Religious Experience.* New York: New American Library (Mentor Books), 1958.

6

IT MATTERS

*Technological progress has merely provided us with
more efficient means for going backwards.*
—Aldous Huxley[1]

One of the primary concerns that stimulated the creation of the
PEAR program was the possibility that human technology might
be vulnerable to inadvertent or intentional disturbances associated with the consciousness of its human operators. Although
our early ancestors may well have attributed anthropomorphic
properties to their tools, for the better part of the period following the Scientific Revolution this had not been an issue that attracted much scholarly attention. By the middle of the 20th century, however, information technology (IT) was well on its way
to becoming a dominant feature of contemporary society, and
anomalous interactions of humans with their information processing machines had graduated from parlor games to legitimate
practical problems.

Over recent decades, IT has continued to evolve and proliferate exponentially, consummating in widespread cultural dependence on a network of hardware, software, and infrastructure of
a scale, sophistication, and sensitivity unimaginable fifty years
earlier. We have witnessed the shift from primitive punch cards
to large-scale VAX machines, through countless generations of
PCs, to laptops and miniaturized hand-held devices, to the exploration of quantum computing facilities. Where we once spoke of
memory capacity in terms of kilobytes, we now refer to terabytes
and petabytes; and we have progressed from regarding our computers as sophisticated calculators, to relying on them as basic research tools, indispensable personal aides, brain interface systems,

archival resources, and virtual doorways to worldwide communications. Our ubiquitous cell phones, text messengers, digital cameras, and personal digital assistants are usually close at hand, secluded in our pockets and purses, held or clamped to our ears, collecting, preparing, and processing information for intimate presentation to our minds on matters of imme-

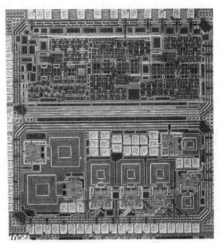

Integrated circuit

diate private relevance, and we Google, Tweet, Friend, Podcast, or blog on a daily or hourly basis. IT today is such an integral and inescapable dimension of our private and professional lives that it now functions as both a tangible and intangible extension of our consciousness. Since virtually all of these devices manipulate strings of binary digits to fulfill their information assignments, and many utilize random sources, our research cautions us that they thereby may have become potential targets for resonant anomalous interactions with, and non-local projections of, the thoughts and feelings of the consciousnesses of their human users.

Over the years that information was emerging as the dominant currency of science and technology, evidence was accumulating that the human/machine anomalies PEAR was studying did not lend themselves well to representation in terms of energy transmission, *per se*. As such, they were more consistent with the hypothesis that these anomalies were embodying slight decreases of entropy—that is, information was being introduced into, or extracted from, the random physical systems addressed, somehow altering the intrinsic binary probabilities involved in their random components. The only discernible correlates of these anomalous effects were primarily subjective in character, such as the human operators' intentions and their degree of emotional resonance with their tasks, yet these subjective aspects were being manifested as *objectively* measurable increments of the output information. While we do not yet fully understand how this subtle form of crosstalk between consciousness and the technical logic and functions of the devices and processes they address may arise, its empirical occurrence nevertheless forces consideration of the extent to which it may compromise their operational integrity, and how this may be precluded, or even beneficially enhanced.

The emergence of information as a third major scientific currency (to supplement tangible substance and all forms of energy), concurrently with the development of increasingly sensitive and complex tools for its extraction and deployment, begs broader questions of how to accommodate inherently subjective but critical factors within a viable scientific paradigm. Canonically it is well established that objective information is to some extent interchangeable with matter and energy, which in turn raises the radical possibility that the subjective aspects of information might also pertain not merely to the perception of physical reality, but even to its actual creation. And since the objective and subjective dimensions of information, like particles and waves in atomic-scale events, appear to be complementary in character, they may entail a degree of uncertainty in their relationship that will need to be acknowledged in particularly sensitive aspects of engineering practice.

Unfortunately, there is also evidence that such crosstalk between consciousness and physical systems is not necessarily always constructive. For example, certain PEAR operators have demonstrated relatively consistent tendencies to produce statistically significant anomalous results *opposite* to their stated intentions. Beyond the laboratory, widespread anecdotal lore and common experiences purport that certain persons are preferentially prone to precipitate random malfunctions of technical equipment; that crucial machinery tends to fail more frequently in situations where it is most urgently needed or where environmental chaos is high; or that physical devices can apparently take on malevolent or benevolent anthropomorphic characteristics. For example, extensive documentation exists of mysterious failures of technical equipment in the presence of certain individuals, a phenomenon sometimes referred to as the "Pauli effect" after the theoretical physicist who was reported to be prone to such events. "Murphy's Laws" comprise a collection of adages expressing the premise that if anything can go wrong, it will. And some test pilots and astronauts will privately acknowledge inexplicable aberrations in their meticulously calibrated guidance, control, and communication systems that during World War II came be labeled as "Gremlin" effects.

Even in more traditional practices, we have long since learned that most sensitive engineering technologies must be physically protected against extremely subtle external influences and internal interferences, and much of this equipment is now routinely shielded from minute thermal, seismic, electromagnetic, and optical disturbances, and even from background cosmic radiation. Facilities for processing delicate biological substances, rare physical materials, or radioactive ingredients may require exceptional environmental controls and discipline of personnel. But in the newly emerging context of anomalous disturbances we are also obliged to consider whether more proactive capacities of human consciousness in its intimate interactions with its technological aids should be taken seriously, and if so, what remedies against its misuse can be invoked. Can we continue to presume the invulnerability of all modern instrumentation, control, and operational equipment to inadvertent or intentional disturbances associated with the psyches of its human operators, especially in periods of intense emotional stress or intellectual demand? Can we be quite sure that the delicate and sophisticated information-processing devices and strategies functioning in juxtaposed microelectronic and mental machinery proceed totally independently, without any interference or crosstalk whatsoever? Or, on the positive side, can we continue blithely to dismiss any possibilities for creating bonded human/machine systems whose capabilities exceed those of their separate human/machine partners?

Regrettably, at this point we are not at all well prepared to protect our contemporary technologies against the possibility of such debilitating negative interaction scenarios, or to develop such potentially beneficial human/machine systems. Further rounds of basic and applied research are needed to clarify, correlate, and quantify such anomalous interaction capabilities, and to control and utilize them. While such studies are relatively simple to pose, they are not so simple to implement. We know that the phenomena involved do not follow the accepted rules of causality, are extraordinarily subtle and elusive, are difficult to replicate, correlate, and interpret, and that few viable theoretical models have yet emerged to facilitate their comprehension. Nonetheless, there is a growing hope that the same increase in the sensitivity and sophistication of equipment that prompts concerns about its vulnerability to such effects may also aid in their resolution. In particular, the very subtle signal levels and very rapid data acquisition and processing capabilities now available may enable experimental access to domains of interaction wherein the phenomena of interest may be more replicable, and systematic study may become more feasible. Also, the scholarly progress in cataloguing, modeling, and understanding the behavior of many other emerging non-linear, complex, and chaotic technological systems may offer valuable metaphoric precedents for representation of various consciousness-correlated physical phenomena. Indeed, these may be our most powerful technical analogies going forward.

Experiments in this context are to some extent related to more conventional engineering research and development that has variously been labeled "human factors," "cybernetics," "ergonomics," or "man/machine interactions," and that has run parallel to similar paths of study in industrial psychology. In fact, the most productive of such efforts have been those that have compounded insights from the engineering and psychological perspectives in a truly interdisciplinary fashion. But any attempt to address anomalous aspects of the interaction of human consciousness with physical devices from these complementary perspectives clearly must contend with a range of relevant subjective issues that remain poorly understood, or cannot even be specified *a priori*. Notwithstanding, if the phenomena derive to any degree from conscious or unconscious processes of the human mind, it is important that the experimental equipment and protocols be designed to enhance, or at least not to inhibit, those subtle effects. Very likely, the difference between sterile experiments and effective ones, or between productive technical applications and unmanageable ones, will lie as much in the impressionistic aspects of their ambience, protocols, and feedback as in the elegance of their instrumentation, and the validity of any theoretical models will depend on the astuteness with which they engage such subjective factors in their mechanics.

It is important that the planning, implementation, and interpretation of this class of research and its pragmatic derivatives incorporate the insights, intuitions, and personal experiences of the human operators who are asked to function as components of the human/machine systems. Given the intrinsically interdisciplinary nature of the topic, it is equally essential that the research personnel be open-minded, congenial, and cognizant of forefront work in relevant disciplines beyond their own. Finally, the institutions housing these efforts need to comprehend and respect these nuances of approach, and display tolerance and support for their unusual tone and special requirements. Nonetheless, the vision of a technology, however subtle and complex, that could reliably sense the degree of coherent purpose and productive resonance possible in such diverse arenas of human dynamics as business and industry, healthcare, education, athletics, artistic performance, and creative scholarship, among countless others, and lead to beneficial applications therein, seems to merit continued exploration.

All of the original and extended conceptual and strategic vectors described in this Section clearly converge on a common epistemological need, namely, a program of modern research that can directly address the role of consciousness in the establishment of physical reality. At their confluence, these diverse perspectives constitute an unresolved chord of enigmatic experience and incompletely substantiated speculation that yearns for definitive evidence and incisive explanation to confirm and clarify its message. It is to this end that the PEAR program described in the following sections has been addressed.

Reference

---

[1] Aldous Leonard Huxley. Interview, *Time*. Monday, Nov. 29, 1937.

SECTION II
Human/Machine Connections:
Thinking Inside the Box?

PEAR REG

1

SCIENCE OF UNCERTAINTY

You think that because you understand ONE
you understand TWO, because one and one makes two.
But you must understand AND.
— Sufi proverb

Throughout its entire thirty-year lifetime, the PEAR program confined itself to three distinct but synergistic sectors of study. The
first, which is the subject of this Section of the book, comprised
an ensemble of experiments investigating anomalous interactions
of human consciousness with various random physical devices,
systems, or processes that resulted in statistical output characteristics deviating significantly from those expected on the basis
of known scientific mechanisms, normal technical behaviors,
or simple chance expectations. The second, covered in Section
III, concerned "remote perception"—the acquisition of information about physical locations remote in distance and time and
inaccessible by any known sensory communication channels.
Technically disparate as these two arenas of empirical study
may appear superficially, the degree of similarity they displayed
in the character and scale of their results, their salient physical
and psychological correlates and lacks thereof, and the subjective
involvements and responses they stimulated, all suggested that
these diverse anomalies actually were drawing from the same,
deep, primordial Source of reality. The third component of the
program, presented in Section IV, was the development of integrative theoretical models that have proved useful for correlating
both classes of experimental data, for designing more incisive
experiments, and for explicating the basic phenomena on fundamental grounds.

The genesis and early course of this conglomerate enterprise, along with the ambience and style of its laboratory operations, have been well represented in *Margins of Reality*,[1] and the technical and analytical details have been covered comprehensively in a large number of referenced journal publications, technical reports, and review documents.[2,3] For those readers who are themselves active in this category of research or who are otherwise well familiar with our program, most of the operational and analytical mechanics of the business are already in place and further attempts to include these elements here seems redundant. Younger scholars aspiring to enter this field, however, or readers new to the topic or seeking a more informed basis for their personal judgments of its credibility, relevance, and potential in their own lives, may find some elementary description of the empirical, analytical, and theoretical techniques involved, as well as their limitations, to be helpful. Yet more casual readers may prefer to move directly to our salient results and their potential implications, unencumbered by the esoteric logic and jargon that lead up to them. Thus, in this Section we shall review only in general terms the principal findings of our human/machine experiments, along with their relevant correlates and significance, to begin construction of a credible platform for the deductive interpretations of these strange behaviors to be proffered in Sections IV and V.

It is important to emphasize at the outset that all of the data acquired throughout this research program are inescapably statistical in character. More bluntly, every individual and collective result achieved in these studies *could* have been a random "chance" event. The only criterion we can apply to discriminate between noteworthy anomalies and sheer coincidence is how *likely* those "chance" occurrences would have been. In other words, such events are not deterministically causal; they are *probabilistic*, and therefore embody varying degrees of *uncertainty*. The philosophical ramifications of this recognition necessarily spread very deeply throughout the epistemological perspectives, extending even to the conceptual definition of "chance" itself. Friedrich Schiller stated his opinion unequivocally:

There is no such thing as chance and what we regard as blind circumstance actually stems from the deepest source of all.[4]

(Schiller employed Old German vocabulary in his writing. He opted for the term "Quelle" which an old German dictionary defines as "outgushing stream, spring, well, source, fountainhead; original source of knowledge, ultimate authority or document."[*]) Samuel Taylor Coleridge subscribed to a similar if somewhat more theistic conviction:

> Chance is but the pseudonyme of God for those particular cases which He does not choose to subscribe openly with His own sign manual.[5]

It is this ineffable "Source", by whatever name, that the research sketched in the following text has inescapably engaged.

A. Statistical Stance

Notwithstanding, we must have in hand some basis for distinguishing "anomalous" behavior from "normal," and in the scientific world, mathematical statistics appears to be the weapon of choice. In the opening Section of *Margins*, we offered a simplified tutorial representation of those tools of elementary statistics that were being applied routinely to quantification of the PEAR empirical evidence, and that of many other laboratories, past and contemporary. With reference to that, and to the many readily available statistical textbooks, our treatment of this analytical aspect here will be limited to passing mention of those techniques that continue to be deployed to assess the statistical merit of the various arrays of such experimental results. (Readers preferring not to involve themselves with this analytical formalism or its associated technical language may safely skip the following few pages and rejoin the narrative at the start of the next subsection.)

---

[*] *Whitney's German-English and English-German Dictionary.* Yale College, New Haven: Henry Holt & Co., 1877.

For virtually all of these bodies of data, simple "parametric" models (*e.g.* those correlating data with specified experimental conditions) more than suffice to represent the scale and character of the putative empirical anomalies. In those rare cases where more sophisticated treatments have been useful, *e.g.* analysis of variance (ANOVA), Bayesian methods, Monte Carlo simulations, etc., these will be specifically noted in due course. Virtually all of these techniques derive from binary combinatorial algebra, which for the very large databases that have prevailed in essentially all of the experiments, converge to excellent approximation to the ubiquitous "Gaussian" distributions, commonly dubbed "bell curves," the details and mathematical treatment of which are routinely central to all elementary statistics.

For most of the experiments described in the following chapters, the theoretically expected chance distributions are adequately represented in the conventional Gaussian algebraic form:

$$f(x) = \frac{1}{\sqrt{2\pi}\sigma} e^{-\frac{1}{2}\left\{\frac{x-\mu}{\sigma}\right\}^2}, \qquad (2.1)$$

where *f(x)* denotes the population density of the counts of value *x* of one of two possible binary digits comprising particular distributions, and the Greek symbols μ and σ denote the means and standard deviations of these theoretical reference distributions. (The square of σ is commonly referred to as the variance of the distribution.) Given their symmetry, the mean values of these distributions are located at their peaks, and their widths are scaled by their standard deviations.

Most of the experiments focus on consciousness-correlated displacements of the binary counts *x*, or of the means of their empirical distributions, *m*, from their expected chance means, μ, and in most cases, the mean shifts ($m - \mu$) are the specified targets of the operator efforts. From these mean shifts can be constructed the commonly utilized "standard scores," or "z-scores," which indicate how many standard deviations a given observation is above or below the mean of its theoretical distribution:

$$z = (x-\mu)/\sigma \qquad\qquad (2.2)$$

In turn, it is readily possible to calculate, via definite integration over the pertinent distributions, associated probabilities, p_z, that represent the likelihoods that particular values of z, *or any larger values of them*, could have occurred by chance. These "tail probabilities" then can be compared with some arbitrary threshold value agreed upon in advance to represent a likely anomalous result of occurrence by chance, say .05, or odds of one in twenty, which conventionally limits a "significant" result, or .01, odds of one in one hundred, a "highly significant" one. In those cases where the theoretical reference distributions are unavailable for any reason, so-called "t-scores" can be calculated using somewhat more elaborate relations utilizing the empirical means (m) and standard deviations (s), and the corresponding probabilities of their chance occurrences can be computed similarly.

A complementary, somewhat more tangible index also frequently invoked is the "effect size" of the anomalous mean shift, typically calculated as:

$$\varepsilon = z/\sqrt{n} = (m - \mu)/(\sigma/\sqrt{n}) \,, \qquad\qquad (2.3)$$

or some similar definition appropriate to the particular experimental design. The measures z and ε complement one another in specifying the magnitude of an anomalous event: the absolute departures of the experimental measurements from their chance expectations are scaled by ε; whereas z indicates the statistical merits of those shifts when sustained over database sizes of n observed units. (The ratio $\sigma/\sqrt{n}$ is sometimes referred to as the "standard error" of the extended distribution at hand.)

Thus, the importance of an empirical anomaly depends on three intertwined factors: the intrinsic scale of the effect (ε); the natural *spread of the measurable* (σ or m); and the persistence of the deviation over an extended database (n). This entanglement will be illustrated more specifically in several chapters to follow.

Another visually instructive tactic is the *cumulative deviation* graph, which concatenates sequentially the compounding deviations from the chance mean as the database of interest grows. In this format it is useful to superimpose for comparison the loci of the limits of the chosen chance probability thresholds. These parabolic envelopes indicate the increasing width of particular confidence intervals about the theoretical mean as the database evolves, as illustrated in Fig. II-1.

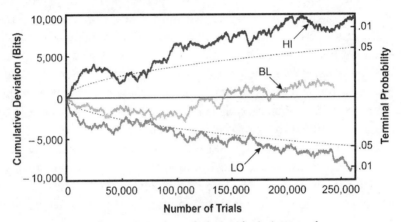

Fig. II-1. Sample cumulative deviation graph.

Such cumulative deviation graphs also can represent characteristic "signatures" of particular operators on particular experiments, by depicting the evolution of their results over extended periods of effort.

In a variety of situations, anomalous behavior may be indicated by other than direct deviations from the theoretical mean. For example, the *shape* of the empirical data distributions may deviate excessively from the idealized theoretical curves. If these distortions are relatively smooth, one may compare a hierarchy of "form factors" or "higher moments" of the empirical distribution functions, such as their "variance," "skew," or "kurtosis," for which corresponding theoretical chance expectations also can be computed, along with appropriate algebraic comparisons to indicate the likelihoods of those disparities. For more severe distortions, one may apply various "goodness-of-fit tests," such as "sum over chi squared" criteria, polynomial regressions, or a variety of other analytical methods. In those situations where the data present anti-symmetric results for opposite directions of intention, other forms of chi-squared analyses, or "Monte Carlo simulations," can uncover additional structural anomalies.

Over the following chapters, we shall endeavor to specify the particular analysis techniques found most effective for given situations. In all of this, however, we must retain our original caveat: In this work, no result is unequivocally anomalous; likewise, none is unequivocally attributable to chance. We only can calculate the *probabilities* thereof, and then specify our own tolerance levels for their degrees of deviation. Thereby, our scientific judgment acquires an inescapably subjective component whose role in the very establishment of the phenomena becomes relevant, but tantalizingly obscure. This issue is addressed more directly in a less widely used statistical technique termed "Bayesian statistics", after its inventor and proponent, the Reverend Thomas Bayes.[6] Although we have employed this method on occasion to confirm results or to respond to criticisms, it has rarely enlightened interpretation of any empirical features.

B. Experimental Strategy

The desirable characteristics for experiments to explore the interaction of human operators with potentially vulnerable physical processes follow directly from various aspects of the phenomenological experience and statistical logic just sketched:

1) The device or system with which the human operators attempt to interact should embody some normally random physical process that can be transcribed into a readily discernible output distribution of data in a form amenable both to direct quantitative recording and to stimulating feedback for the operators.

2) The statistical character of this output distribution should be simple, direct and, ideally, theoretically calculable.

3) The experimental equipment should permit very high signal-to-noise discrimination and provide extreme protection against technical malfunctions, environmental artifacts, operator mishandling, or other illegitimate influences.

4) The entire experimental system should be capable of rapid acquisition of very large databases.

5) Extensive and regular calibrations should confirm the stability of the system and its continuing unattended conformity to theoretical expectations.

6) The data collection, storage, and computation facilities should provide redundancy of records and be capable of rapid and precise rendering of experimental data into instructive quantitative representations of results.

7) Experimental protocols should optimize operator incentive and ease of operation, emphasize the primary parameters of interest, and provide further protection against spurious results.

8) Beyond its technical sophistication, the physical and social ambience of the experimental facility should be conducive to uninhibited participation of the operators in both the performance and interpretation of the experiments.

In any practical implementations, various optimizations and trade-offs among these desirable characteristics inevitably must be faced, but to be both effective and credible in this type of experimentation, the composite system must respect both the subjective and objective requisites of the interactions.

While this text holds no aspirations to any sort of encyclopedic review of the many designs, applications, and results of the myriad of REG experiments that have been conducted in this field at many laboratories worldwide, we would be remiss in not acknowledging the pioneering and seminal work of several cogent predecessors. In particular, the work of Helmut Schmidt[7-9] and Dean Radin,[10-12] among many others, has laid a valuable foundation for development of equipment, protocols, and inter-pretations that we have built upon in our subsequent studies, to the extent that ours might better be regarded as replications and extrapolations of their earlier and on-going efforts.

The most popular technical sources for provision of such output data distributions have been electronic Random Event Generators (REGs), various forms of which have served as the basis for most of our own human/machine studies, and those of many other laboratories. Typically, such devices employ a source of physical "white noise" generated by some random microscopic physical process, *e.g.*, a thermal electron current, a gaseous dis-charge, a photon emitter, a radioactive decay, *et al.* Logic circuitry transforms this microscopic noise into a regularly spaced string of randomly alternating binary pulses (bits), which subsequent cir-cuitry counts, displays, and records. All of this can be performed with great precision and safeguard against artifact, at very rapid rates, constituting a genre of prolific electronic "coin-flipping" devices toward which the operators may direct their intentions.

Various versions of such microelectronic sources have been used in our experiments, along with an assortment of mechanical, optical, fluid-dynamical, and acoustical devices, to be described in subsequent chapters of this Section. The remarkably similar char-acteristics of the effects achieved across this disparate range of sources must be regarded as salient indicators of the fundamental nature of the phenomena that will need to be conceded in any viable theoretical models.

References

[1] Robert G. Jahn and Brenda J. Dunne. *Margins of Reality: The Role of Consciousness in the Physical World.* San Diego, New York, London: Harcourt Brace Jovanovich, 1987. Reprinted: Princeton, NJ: ICRL Press, 2009.

[2] Robert G. Jahn and Brenda J. Dunne. "The PEAR proposition." *Journal of Scientific Exploration, 19,* No.2 (2005). pp. 195–246.

[3] Robert G. Jahn, Brenda J. Dunne, Roger D. Nelson, York H. Dobyns, and G. Johnston Bradish. "Correlations of random binary sequences with pre-stated operator intention: A review of a 12-year program." *Journal of Scientific Exploration, 11,* No. 3 (1997). pp. 345–367.

[4] Johann Christoph Friedrich von Schiller. *Wallensteins Tod* (The Death of Wallenstein), Act II, Scene 3. W. Witte, ed. Oxford: Basil Blackwell (1952). p. 186.

[5] Samuel Austin Allibone. *Prose Quotations from Socrates to Macaulay.* Philadelphia: J. B. Lippincott & Co., 1876. p. 91.

[6] Thomas Bayes. "An Essay toward Solving a Problem in the Doctrine of Chances." *Philosophical Transactions of the Royal Society of London,* Vol. 53 (1763), pp. 374–418.

[7] Helmut Schmidt. "A PK test with electronic equipment." *Journal of Parapsychology, 34* (1970). pp. 175–181.

[8] Helmut Schmidt. "PK tests with a high-speed random number generator." *Journal of Parapsychology, 37* (1973). pp. 105–118.

[9] Helmut Schmidt. "PK effect on pre-recorded targets." *Journal of the American Society for Psychical Research, 70* (1976). pp. 267–291.

[10] Dean I. Radin. *The Conscious Universe: The Scientific Truth of Psychic Phenomena.* San Francisco: HarperEdge, an imprint of HarperSanFrancisco, 1997.

[11] Dean I. Radin. *Entangled Minds: Extrasensory Experiences in a Quantum Reality.* New York: Simon & Schuster, 2006.

[12] Dean I. Radin and Roger D. Nelson. "Evidence for consciousness-related anomalies in random physical systems." *Foundations of Physics, 19,* No. 12 (December 1989). pp. 1499–1514.

2

BENCHMARK BEGINNINGS

To attain the impossible one must attempt the absurd.
—Cervantes[1]

In the earliest days of the PEAR program, online computer access was not an option. Rather, REG data were recorded on a strip chart printer and subsequently entered by hand into a mainframe computer; the initial statistical calculations were performed manually. The first revelation of the character of the anomalous effects we would be dealing with emerged from an intensive session of hand-plotting the individual counts of one of our early operator's first 5,000 experimental trials. Using red dots to graph the counts from the HI-intention efforts, and green dots for the LO, it soon became evident from the emerging patterns that we were looking at two distinct distributions whose separation had a statistical likelihood of a few parts in ten million.

"That's very nice," observed Bob, as he put down his colored pencils with a mild grunt of satisfaction.

"What do you mean, 'very nice'?" demanded Brenda. "That's absolutely incredible!"

From either perspective, it was clear that these results merited further study, and this particular dataset served as the original component of what we came to refer to as our Benchmark REG experiment.

As the world of information technology evolved, our data processing became substantially more sophisticated and efficient, and the microelectronic REG became the workhorse of the laboratory, setting a precedent for many subsequent experiments. Its capacity for rapid trial generation, attractive feedback, and eventual online recording and computerized statistical analyses, permitted

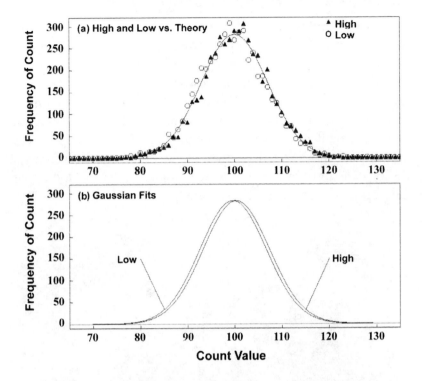

Fig. II-2. Count distributions of one operator's first 5000 REG trials superimposed on theoretical chance expectation: a) HI- and LO-intention data; b) Best binomial fits to HI and LO data.

the accumulation of enormous quantities of data in reasonably short periods of time, and the experimental protocols proved sufficiently appealing to our operators that they were willing to commit themselves to extended periods of involvement. While our technology and protocols underwent further modifications over time, the essential tri-polar design, where participants generated trials under three conditions of intention, remained constant throughout.

A. Experimental Equipment
The particular REG apparatus first regularly deployed in our experiments utilized as its random source a commercial microelectronic

noise diode unit commonly incorporated in a variety of communications, control, and data-processing equipment. In addition to the filtering, sampling, counting, and display circuitry that rendered this noise into an output distribution of binary counts, the system entailed extensive fail-safe and calibration components that guaranteed its integrity against technical malfunctions and environmental disturbances. A photograph of an operator interacting with the device is shown, along with a block diagram of its logic circuitry in Fig. II-3. More complete details of the design and operation of this machine have been presented in Refs. 2–5.

Benchmark REG with operator

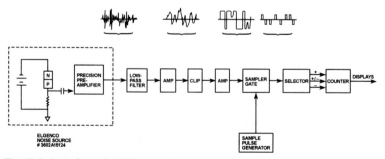

Fig. II-3. Benchmark REG function diagram.

This REG constructed its binary output string by sampling the conditioned electronic noise pattern at preset regular intervals. If the noise signal happened to be positive (greater than its mean value) at the time of sampling, a positive pulse (bit) was produced; if the noise signal was negative at that time, a negative bit was produced. The basic experimental data group, called a trial, comprised some given number of these electronic bits. By dial setting, the machine could be instructed to form 20, 100, 200, 1,000, or 2,000 bits for each trial. Similarly, it could be told how rapidly to take the samples (10, 100, 1,000, or 10,000 times per second) and whether to count only positive bits, only negative bits, or to count those bits conforming to a regular alternation (+, −, +, −, +, −, ...). This last, alternating mode served to eliminate any effects of residual bias in the basic noise pattern and was employed for virtually all data reported herein.

B. Protocol Possibilities

The primary goal of most of the early REG studies was to explore correlations of statistical shifts in the output count distributions with pre-stated intentions of the operators. To establish any such alterations in the most incontrovertible terms, the experimental protocol encouraged the accumulation of large blocks of data wherein all controllable secondary parameters were held fixed, and the primary variable—the pre-recorded intentions of the operator—was explored in a tripolar protocol. Data generated under operator intentions to achieve higher than chance trial counts (HI), lower than chance counts (LO), and "baseline" data taken with the operator present but with no conscious intention (BL), were interspersed in some balanced recipe. Only if systematic deviations of significant statistical magnitude were observed among the ensuing three streams of data were correlations of the results with operator intention claimed. In this manner, any artifacts introduced by the equipment or environment could not affect the results over large blocks of data, unless they themselves were somehow correlated with operator intention.

An important strategic option was the number of trials that were blocked together during acquisition and processing of the data. In essence, these block sizes were set by a compromise between the establishment of clear systematic trends in the data, and the stamina and attention span of the operators. Largely on the basis of empirical experience with both of these factors, three levels of data agglomeration were maintained. The smallest of these, called a *run*, comprised a sequence of trials strung together under a single intention of the operator, without pause for manual recording or other break in attention. The bulk of our early REG data was acquired in runs of 50 or 100 trials, although a substantial base of data was also generated in 1000-trial runs. Once the run length was selected, it remained constant throughout a given experimental series.

A second, more flexible grouping of trials was termed a *session* and consisted of the number of interspersed HI, LO, and BL runs the operator completed in any one period of access to the equipment. The choice of session length was left largely to the preference of the operator, although certain minimum block requirements were imposed, depending on the prevailing protocol being followed, to preclude optional stopping effects from distorting the data.

The largest and most important unit of data concatenation was the *series*, which typically consisted of 1,000, 2,500, or 5,000 trials in blocks of 50 or 100 runs, or 3,000 trials in blocks of three 1000-trial runs, under each of the three intentions, HI, LO, and BL. The series length also was fixed at the start of an experiment, but the amount of time required varied considerably from operator to operator. Some took two to six weeks to complete a full tripolar series, others as little as one day or as long as a year. Despite the major commitment required of an operator to generate such a series, we found this to be the minimum base of data from which consequential systematic trends could reliably be extracted from the inherently statistical variations, and therefore most of our statistical assessments were performed on this scale of data.

Beyond the primary variable of operator intention, a number of secondary parameters suggested themselves for systematic experimental survey. For example, the various technical options provided by the machine for sampling rates, sample sizes, or counting of only positive or only negative bits could conceivably have had psychological implications for the operator's performance of the primary task, or even some fundamental relevance to the nature of the anomalous interaction itself, and attracted considerable exploration. Even more instructive were differences in the performance of the operators when allowed to choose the directions of intention versus when assigned the direction by other persons or random processes. In the *volitional* mode, the operator decided in advance the sequences for interspersing HI, LO, and BL runs, and pre-recorded them in the database. In the *instructed* mode, the operator activated an ancillary random criterion in the REG before each run that specified in advance the direction of intention and automatically recorded the assignment.

A similar division of protocol explored the importance of the control by the operator of the time of initiation of each trial. In the *manual* mode, the operator began each trial by a discretionary button push; in the *automatic* mode, a regularly repeated sequence of 50, 100, or 1,000 trials was initiated by a single button push.

To recapitulate the ranges of experimental units and secondary parameters explored in these microelectronic REG experiments:

Experimental Units:
- Trial: Number of electronic bits processed (20, 100, 200, 1,000, 2,000)
- Run: Number of trials in single operator effort (50, 100, 1,000)
- Session: Number of runs in one period of experimental effort (Operator preference)
- Series: Data blocks for statistical processing (1,000, 2,500, 3,000, 5,000 trials per intention)

<u>*Secondary Parameters:*</u>
- Instructed/Volitional assignment of intention
- Automatic/Manual sequencing of trials
- Number of bits per trial
- Sampling Rate (10, 100, 1,000, 10,000 bits per second)
- Feedback Format (LED display, computer displays, none)
- Number of co-operators

In a typical experiment, operators first selected a set of the options from those listed in the table, recorded their choices in the logbook and on the computer, and set the REG controls and computer indices accordingly. If the experiment was to be volitional, they then decided their direction of effort and recorded it; if instructed, they activated the random instruction program and acknowledged its result. Then, seated a few feet from the machine, they initiated its operation by a button push or remote switch and endeavored to influence its output to conform to their stated intentions, by whatever strategy they felt effective. Baseline runs were interspersed in some reasonable fashion in appropriate numbers to balance the intentional trials.

These experiments proceeded within a homey decor of paneled walls, carpeting, and comfortable furniture, with options for lighting, background music, or snacks. The laboratory staff maintained a minimal presence during the actual operations, but openly described the experimental goals, gear, and protocols prior or subsequent to the sessions in as much detail as requested. They displayed cheerful, albeit professional attitudes of openness to any results or lack thereof, and encouraged the operators to take a playful approach to the task, rather than being concerned about the outcome. The computer software for specifying, initiating, and recording experimental runs was designed to be user friendly and to encumber the operator as little as possible with operational tasks.

Operators were encouraged to engage in the experiments whenever they were in the mood, or when they wished to explore correlation of their performance with particular attitudes or

PEAR reception area

A PEAR experiment room

emotional states. As mentioned, the length of a session was left to operator preference and varied considerably. Using the automatic mode at a counting rate of 1,000 bits per second, most operators could complete a session consisting of 1,000 trials per intention in less than one hour. Experiments using slower counting rates, larger sample sizes, longer run lengths, or manual operation were correspondingly longer.

C. Data Depots

Quantitative evaluation of experimental data generated in this fashion needed to be consistent with the qualitative aspects of the results. In most cases, the most reliable indicators of the statistical character of the data were the distributions of trial scores obtained over full experimental series. If these distributions closely resembled those of the calibrations and theoretical expectations, then simple statistical indicators could be applied in their analysis. If more complex distortions were evident, more sophisticated treatments would be required.

Over the first 12-year period of experimentation, 91 individual operators, all anonymous, uncompensated adults, none of whom claimed unusual abilities, accumulated a total of 2,497,200 trials, each of 200 binary samples, most counted at a rate of 1,000 per second, and distributed over 522 tripolar series. Hereinafter, this body of data will be referred to as our "Benchmark" database, which actually comprised three separate components, all following the same basic protocol, but exploring an assortment of secondary experimental parameters. The earliest phase, designated "OldREG," consisted of 120 series comprising a total of 345,000 trials per intention, with each series consisting of 5,000 trials generated in 50-trial runs.

The second phase, labeled "RemREG," was designed to assess the dependence on physical separation of the operator from the machine. This was pursued by generating data under three conditions: a) with the operator present in the REG room; b) with the operator personally initiating the runs, but sitting in an adjoining room while the device was running; and c) with the operator in a remote location, with the runs initiated by a staff member who had no knowledge of the operator's intentions until after the data had been recorded. Each of these components of RemREG consisted of 3,000 trials per series, generated in runs of 1000 trials per intention, and yielded a total of 69 series (or 207,000 trials per intention) in the proximate condition, 17 series (51,000 trials per intention) in the "next room" condition, and 106 series (318,000 trials per intention) in the "remote" condition.

The third phase, termed "ThouREG," consisted of 508 series of 1,000 trials per intention, generated in runs of 50, 100, or 1,000 trials, chosen in accordance with operator preference, and totaled 508,000 trials per intention.

Since in all three of these Benchmark phases data were generated in trials of 200 bits, with all secondary parameters remaining constant throughout a given series, it was legitimate to combine all of the local data into a single large database for composite analyses. Table II-1 lists the overall results of this composite Benchmark database for the three categories of intention (HI, LO, and BL) and for the HI − LO separations, to be compared with concomitant calibration data and theoretical chance expectations. With reference to the Table Key, the salient indicators are the mean shifts from the theoretical expectation, Δ_μ, the corresponding z-scores, z_μ, and the one-tailed probabilities of chance occurrence of these or larger deviations, p_μ. Also listed are the proportions of the 522 local series yielding results in the intended directions (S.I.D.), and the proportions of operators achieving results in the intended directions (O.I.D.).

These measures individually and collectively serve to define the scale and character of the primary anomaly observed in these studies, *i.e.*, the statistically significant correlations of the output of this microelectronic random binary process with the pre-stated intentions of a large pool of unselected human operators. Specifically to be noted is the overall scale of the effect, ε_μ, on the order of 10^{-4} bits inverted per bit processed; the somewhat higher deviation in the HI results compared to the LO; the slight departure of the BL results from both the theoretical chance expectation and the calibration value; and the negligible alterations in the variances of the score distributions, which allow the substitutions of z-scores for *t*-scores. The overall figure of merit for the HI − LO separation, which was the postulated primary indicator, was $z_\mu = 3.81$ ($p_\mu = 7 \times 10^{-5}$). The anomalous correlations also manifested in the fraction of experimental series in which the terminal results confirmed the intended directions. For example, 57% of the series displayed HI − LO score separations in

TABLE II-1

Statistical Measures from All Local Benchmark REG Experiments

Measure	CAL	HI	LO	BL	HI − LO
N	5,803,354	839,800	836,650	820,750	1,676,450
m	99.998	100.026	99.984	100.013	0.042
σ	7.075	7.070	7.069	7.074	—
$\Delta\mu$	−0.002	0.026	−0.016	0.013	0.042
$\varepsilon_\mu \times 10^5$	−1.000	13.000	−8.000	6.500	21.000
z_μ	−0.826	3.369	−2.016	1.713	3.809
p_μ	0.409	3.77×10^{-4}	0.0219	0.0867	6.99×10^{-5}
S.I.D.	—	0.52	0.54	0.50	0.57
O.I.D.	—	0.62	0.47	0.59	0.52

KEY

N: Number of trials performed (200 binary samples each);
m: Empirical mean of trial score distribution;
σ: Standard deviation of trial score distribution;
$\Delta\mu$: Difference of mean from theoretical chance expectation, $m - \mu$;
ε_μ: Effect size, computed as $\Delta\mu/2\mu$;
z_μ: z-score of mean shift, $= \sqrt{N}\,\Delta\mu/\sigma$;
p_μ: One-tailed probability of z_μ (BL is treated as in intended direction when positive);
S.I.D.: Proportion of series having z_μ in the intended direction;
O.I.D.: Proportion of operators with overall results in the intended direction.

---

the intended direction. (The corresponding z-score and associated chance probability for this series-result separation are z_s = 3.15, p_s = 8 × 10⁻⁴).

Alternative display of these overall long-term results in the form of cumulative deviation graphs was also instructive (Fig. II-4). In this format, the anomalous trends in the HI and LO performances appeared as essentially random walks about shifted mean values, leading to steadily increasing departures from expectation. Consistent with the terminal values listed in Table II-1, the average linear fits to the slopes of these two patterns of achievement, in units of bits deviation per bit processed, were roughly 1.3 × 10⁻⁴ and −7.8 × 10⁻⁵ respectively. Although the individual operator performances varied considerably from one another, and from time to time were themselves inconsistent, their compounded effects displayed trends that deviated strongly from chance expectation, with probabilities of 4 in 10,000 for the HI efforts and 2 in 100 for the LO. The chance likelihood of the HI – LO separation was 7 in 100,000.

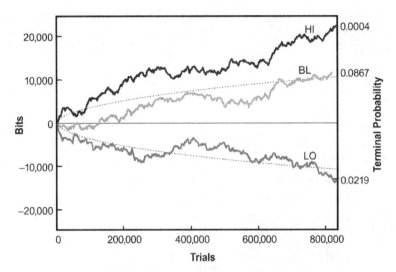

Fig. II-4. Cumulative deviation graphs of composite local Benchmark REG results for HI, LO, and BL operator intentions. Parabolic envelopes are one-tailed 95% confidence intervals about the theoretical chance mean. The scale on the right ordinate refers to the terminal chance probabilities.

In passing, three observations should be noted regarding the drift of the baseline data away from the theoretical chance mean reference line toward slight upward penetration of the positive 95% confidence parabola. First, the baseline condition, where the operator is present but not expressing a conscious intention, should be distinguished from calibrations, which were generated automatically with no one present and fell much closer to the theoretical mean. Second, since baseline trials by definition entail no preferred direction, a two-tailed, rather than one-tailed significance criterion is appropriate, for which the corresponding parabolic envelope ($z = \pm 1.96$) would be wider than that sketched on the figure. Third, it later was found that the major portion of this baseline drift was contributed by the female operators, a feature which holds considerable theoretical implications, as will be discussed further in Chapter II-13, and in Sections IV and V.

These Benchmark results largely defined a hierarchy of questions which guided the subsequent course of our experimental program, *e.g.*:

- With what consistency and over what duration could a given operator produce statistically anomalous effects?
- To what extent did the results depend on the various operational parameters?
- To what extent did the results depend on the particular noise source or logic circuitry utilized in the REG?
- To what extent did the results depend on psychological or physiological characteristics of the operators?
- How did combined efforts of more than one operator reinforce or detract from one another?
- How did the results depend on the physical separation of the operators from the machines, or on the temporal separation of operator efforts from the time of machine operations?
- Did the details of the anomalous output distributions provide any hints about the essential origin of the phenomena?

Definitive pursuit of each of these structural aspects required a lengthy, dedicated, empirical study in its own right, compounding to a constellation of efforts to dissect the complex of parametric interdependencies that contribute to the composite phenomenon. Much of the following text in this Section describes this panoply of studies that attempted to flesh out our skeletonic understanding of this intricate entanglement of human/machine anomalies.

References

[1] Miguel de Cervantes Saavedra. *Don Quixote*. New York: Echo, 2003.

[2] Robert G. Jahn and Brenda J. Dunne. *Margins of Reality: The Role of Consciousness in the Physical World*. San Diego, New York, London: Harcourt Brace Jovanovich, 1987. Reprinted: Princeton, NJ: ICRL Press, 2009.

[3] Robert G. Jahn, Brenda J. Dunne, Roger D. Nelson, York H. Dobyns, and G. Johnston Bradish. "Correlations of random binary sequences with pre-stated operator intention: A review of a 12-year program." *Journal of Scientific Exploration, 11*, No. 3 (1997). pp. 345–367.

[4] Robert G. Jahn, Brenda J. Dunne, and Roger D. Nelson. "Engineering anomalies research." *Journal of Scientific Exploration, 1*, No. 1 (1987). pp. 21–50.

[5] Robert G. Jahn and Brenda J. Dunne. "The PEAR proposition." *Journal of Scientific Exploration, 19*, No. 2 (2005). pp. 195–246.

3

THE DEVIL IN THE DETAILS

...when you have eliminated the impossible
whatever remains, however improbable,
must be the truth.
— Sir Arthur Conan Doyle[1]

Most of the parametric queries raised at the close of the preceding chapter were in fact pursued empirically and analytically to the extent that they provided useful insights into various aspects of the phenomena, and responses to these can now be given with some authority of data; the remainder still merit further study. In conjunction with these *ad hoc* explorations, an assortment of retrospective searches for structural patterns within the extant Benchmark databases were also undertaken, several of which yielded informative indications pertinent to the construction of theoretical models, as developed in this and subsequent chapters. To summarize briefly, qualitative inspection of these data, supplemented by comprehensive analyses of variance,[2,3] indicated that most of these factors did not consistently alter the character or scale of the overall operator effects from those revealed in the Benchmark data, although some might have contributed to certain individual operator performance patterns. However, a few instructive details could be extracted from these surveys.

A. Operator Consistency

The extent of operator replicability that could be achieved can be illustrated by the results of one prolific operator who, over the course of six years of Benchmark effort, produced a total of 15 series. Figure II-5(a,b) compares the cumulative deviation results of the first series (5,000 trials/intention), with those achieved over all 15 series (approximately 55,000 trials/intention).

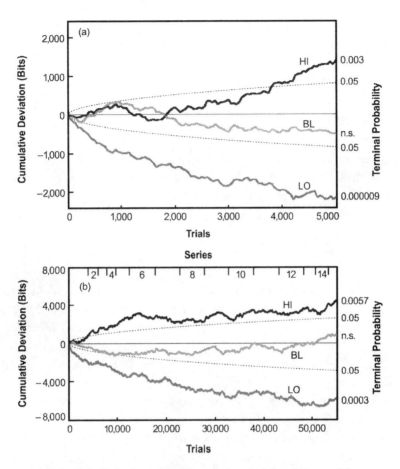

Fig. II-5. REG data of one prolific operator: a) Deviations of first series (5000 trials); b) Deviations of 15 series (55,000 trials).

Clearly, the overall statistical character of this operator's performance persisted from series to series, as well as within each series, despite the inevitable span of psychological and environmental conditions that must have been subsumed over so long a time and so immense a database. To this extent then, the effect was statistically replicable, and the pattern could be regarded as a "signature" of this operator on this particular experiment. Several other operators were found to display similar consistencies of performance, appropriately scaled to their individual database sizes; yet others displayed less replication.[4]

B. Secondary Parameters

The effects on performance of most of the secondary operational parameters, such as volitional vs. instructed assignment, manual vs. automatic trial generation, or feedback mode, likewise were found to be strongly operator specific. That is, for some operators these variables made no noticeable difference; for others, they had significant impacts. As one example, Figure II-6(a,b,c) contrasts the results of one operator working in the instructed mode, wherein the direction of intention was assigned by a random process, with the comparable body of data taken in the volitional mode, wherein this operator was allowed to choose the intended direction for each block of five runs. In the instructed condition, the operator showed a consistent "psi-miss" pattern, inverting the performance from the intention in both directions (Fig. II-6a). In the volitional condition, however, the performance was well correlated with intention in both directions, compounding to a significant positive split in the terminal scores (Fig. II-6b). Note that the scales of achievement and the general patterns of the two signatures were virtually asymmetric, combining to an apparently null overall effect for the total database (Fig. II-6c). (This is an example of the utility of chi-squared tests to supplement the prescribed HI – LO all-data mean-shift calculations. In this case, the latter returned a non-significant result; the former, which gave balanced credit to the separate HI and LO results under each protocol, revealed a clearly anomalous structural feature.)

Such performance disparity suggested that some aspects of the operator psyche may have been resistant to the prevailing constraints, *e.g.,* the imposition of the direction of effort in the instructed protocol, or the unrelenting presentation of new trials in the automatic sequence. However, for other operators just the reverse patterns were observed: strong correlations on instructed or automatic modes and lack of correlation on volitional or manual, from which one might conclude that those particular

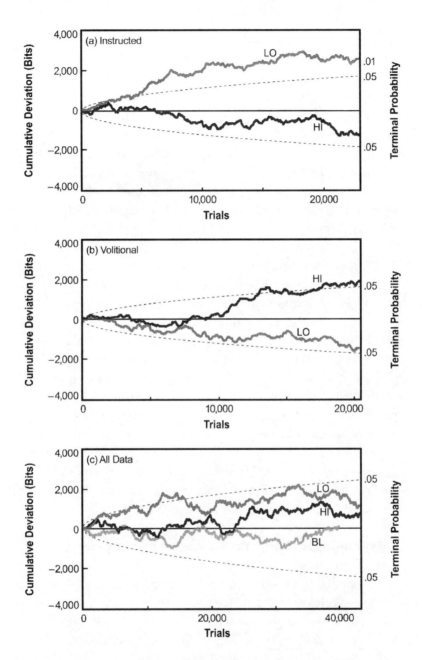

Fig. II-6(a–c). Sensitivity to instructed/volitional option for one operator: a) Instructed; b) Volitional; c) All data.

operators preferred not to choose their directions of effort, or the times to initiate each trial. There were yet other examples where such options seemed to have little effect on the operator performance signatures. In sum, it appeared that the influence of these parameters was neither qualitatively nor quantitatively uniform. For some operators they were unimportant; for others they were consequential, but in different ways and to different degrees.

A similar pattern of individual dependence seemed to prevail for other technical parameters such as the number of bits per trial, the sampling rate, or the number of trials per run. Although the multi-varied options for each of these precluded any complete survey of all possible permutations, selective explorations by several of our operators indicated once again that any given parameter might or might not consistently influence any individual signature, but no generic pattern across these operators and experiments was found. On the one hand, this lack of any general correlations of technical options with the observed anomalies complicated considerably the design of experiments and the search for comprehensive explanation. On the other hand, the demonstrated specificity of influence of these parameters on some individual operators raised intriguing possibilities of their eventual correlation with some psychological or physiological characteristics. In a sense, the individual signatures and their dependence on such parameters might constitute some sort of complex and subtle psychic "Rorschach" test that could reveal features of the operator's consciousness not addressed in other venues.

C. Psychological and Physiological Parameters

Although no systematic assessment of any of the multitude of other potentially relevant subjective correlates was attempted, on the basis of the informal discussions with operators, casual observations of their styles, occasional remarks they recorded in the experimental logbooks, their preferences for particular feedback modes, and our own experiences as operators, it was clear that individual strategies varied widely. Most simply attended to the task in a quiet, straightforward manner. A few used meditation or visualization techniques or attempted to identify with the device or process in some transpersonal style; others employed more assertive or competitive strategies. Some concentrated intently on the process; others were more passive, maintaining only diffuse attention to the machine while diverting their immediate focus to some other activity, such as glancing through a magazine, listening to music, or even eating lunch. One of the more intriguing comments in the experimental logbook was from an operator who reported that "Vanilla yogurt really works." Again, little overall pattern of correlation of such strategies with achievement was found. Rather, the effectiveness of any particular operational style appeared to be operator-specific and transitory; what seemed to help one operator did not appeal to another, and what seemed to help a given operator on one occasion might fail on the next. If there was any commonality apparent in this diversity of correlations, it was that the most effective operators tended to speak of the devices in frankly anthropomorphic terms, and to associate successful performances with the establishment of some form of bond or resonance with the device, akin to that one might feel for one's car, tools, musical instruments, or sports equipment.

The only physiological variable we explored in any depth was operator gender, and here we found consistent and significant correlations in virtually all of our successful experiments. This topic will be described in detail in a separate Chapter, as will the questions of combined efforts of more than one operator working in concert, or the effects (or lack thereof) of physical and temporal separation of the operators from the machines.

D. Count Population Patterns

The Benchmark database was also reviewed for evidence of any structural details of the trial count distributions that might illuminate the mechanics of how these were compounding to the observed overall anomalous mean shifts. While no statistically significant departures of their variance, skew, kurtosis, or other higher moments from the appropriate chance values appeared in the overall data, other regular patterns of certain finer scale features could in fact be discerned. For example, with reference to Figure II-7(a,b) and its captions, the proportional deviations of the individual trial count populations from their theoretical chance values were found to scale linearly with their separations from the theoretical mean count.

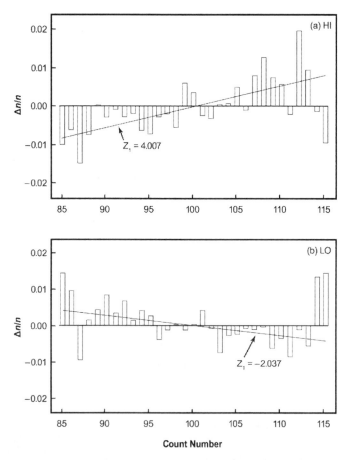

Fig. II-7. Fractional deviations of local count populations from chance expectations accumulated over 91 Benchmark REG experiments: a) HI data; b) LO data. Superimposed straight lines are best-fit linear regressions of statistical significance denoted by attached z-scores. (No higher-order regression terms are significant.)

Such functional behavior can be shown to be consistent with a simple displacement of the chance Gaussian distribution to the overall observed mean values or, equivalently, to shifts in the elementary binomial probabilities from the exact theoretical value of 0.5, to 0.5 plus the pertinent effect size, ε_μ. Given the consistency of all other features of the distributions with chance expectation, this suggests that the most parsimonious model of the anomalous correlations is between operator intentions and the binary probabilities intrinsic to the experiments, *i.e.* that the operator's influence is directed to those binary probabilities, rather than to the establishment of some more complex dynamical mechanism for distortion of the results.[5]

Searches of the data for operator-specific features that might establish some pattern of individual contributions to the overall results were complicated by the small signal-to-noise ratios of the raw data, and by the unavoidably wide disparity among the operator database sizes, leaving graphical and analytical representations of the distribution of operator effects only marginally enlightening. For example, Figure II-8 deploys the 91 individual operator HI − LO mean shift separations as a function of their various database sizes. Superimposed are the theoretical mean value, the mean value of the composite data, and the $p_\mu = 0.05$ deviation loci with respect to these two means. Of primary interest here are the imbalance of the number of positive and negative points about the theoretical and empirical means, the dependence of their statistical significance on their respective database sizes, and the positions of the outliers.

Given the specification of the experimental series as the pre-established unit for data interpretation, and the significantly larger fraction of series having HI − LO differences in the intended directions (*cf.* Table II-1 in the previous chapter), it was also reasonable to search for displacements in the distributions of series scores achieved by all operators. Figure II-9 clearly confirms the overall shifts of the mean, and emphasizes the collective nature of the anomalous effect, vis-à-vis any outlier or "superstar" dominances.

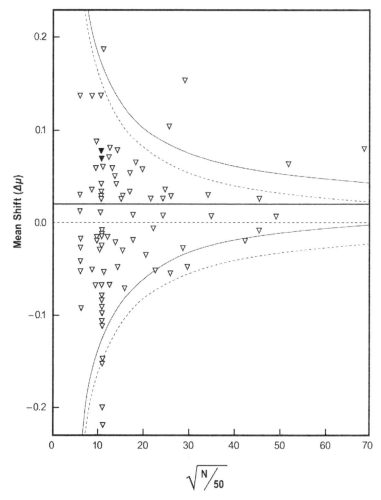

Fig. II-8. Distribution of 91 individual operator HI – LO mean shift separations in the local Benchmark REG experiments, as a function of (re-scaled) database size. Solid triangles denote full overlap of two data points; N = number of trials; .05 probability envelopes superimposed.

E. Learning and Replicability

While it might be reasonable to expect that operators' proficiency at these experimental tasks would improve with increasing experience, no systematic learning trends were evident in the data. Rather, the progression of the anomalous effect sizes as a function

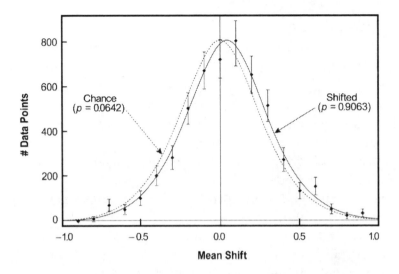

Fig. II-9. Density of data points of 522 local Benchmark series HI – LO mean shifts superimposed on theoretical distributions centered on the chance and empirical mean-shift values. Attached probabilities denote qualities of fits to the two models.

of the number of series completed by the operators was found to take a somewhat unanticipated form. When the HI – LO mean shifts obtained by all operators who had produced at least five series were examined on a series-by-series basis, and these were plotted against that series' ordinal position, a systematic peak of initial success was found to be followed by sharp reductions on the second and third series, after which the effect gradually recovered asymptotically to an intermediate value over the higher series numbers. This pattern obtained, with minor disparities, for the separate HI and LO data, but not for the baselines.[7] It also appeared in a majority of the individual operator databases having five or more series. The interpretation of this pattern on psychological or physical grounds remains somewhat speculative, but its ubiquitous appearance clearly complicates specification of any consistency or replicability criteria. This issue will be revisited in more detail in that context in Chapter 14.

F. Distance and Time

As described in the previous chapter, the RemREG excursions carried out in the second phase of the Benchmark program were designed to explore the dependence of the effect sizes on the distance of the operator from the machine, a factor that could be an important indicator of the fundamental mechanism. Remarkably, no such dependence was found, up to global distances of several thousand miles. In one subset of this "remote" database, the operators addressed their attention to the machine operation at times other than those at which the data were actually generated. Such "off-time" experiments also continued to display a scale and character of anomalous results similar to those of the locally generated data, including gender effects and count population distortions, and as with the spatial separations, no dependence of the yield on the magnitude of the time differences was observed over the range tested. In fact, the overall mean shift in the HI-intention efforts in these "off-time" remote experiments was actually twice as large as that in the "on-time" remote data, although this difference was not statistically significant, given the smaller size of the off-time database. This apparent insensitivity of the anomalies to distance or time was of sufficient theoretical relevance that we continued to explore it in many of our other experiments as well. These will be discussed in more detail in a later Chapter.

G. *Ad Hoc* Additions

While the Benchmark database was the largest single accumulation of REG data acquired at PEAR, many other REG-based experiments were also conducted during the later years of the program to investigate in *ad hoc* fashion a variety of factors that might have further bearing on the nature of the anomalous phenomenon. Several were simple variants of the Benchmark protocol; others were the more extensive excursions described in the following chapters.

References

[1] Sir Arthur Conan Doyle. *The Sign of the Four.* Fairfield, Iowa: 1st World Library — Literary Society, 2004. p. 57. <www.firstworldlibrary.org>

[2] Roger D. Nelson, Robert G. Jahn, York H. Dobyns, and Brenda J. Dunne. "Contributions to Variance in REG Experiments: ANOVA Models and Specialized Subsidiary Analyses." Technical Note PEAR 99002. Princeton Engineering Anomalies Research, Princeton University, School of Engineering/Applied Science, Princeton, NJ. February 1999.

[3] Roger D. Nelson, York H. Dobyns, Brenda J. Dunne, and Robert G. Jahn. "Analysis of Variance of REG Experiments: Operator Intention, Secondary Parameters, Database Structure." Technical Report PEAR 91004. Princeton Engineering Anomalies Research, Princeton University, School of Engineering/Applied Science, Princeton, NJ. November 1991.

[4] Brenda J. Dunne, Roger D. Nelson, and York H. Dobyns. "Individual Operator Contributions in Large Data Base Anomalies Experiments." Technical Note PEAR 88002. Princeton Engineering Anomalies Research, Princeton University, School of Engineering/Applied Science, Princeton, NJ. July 1988.

[5] Robert G. Jahn, York H. Dobyns, and Brenda J. Dunne. "Count population profiles in engineering anomalies experiments." *Journal of Scientific Exploration, 5,* No. 2 (1991). pp. 205–232.

[6] Robert G. Jahn, Brenda J. Dunne, Roger D. Nelson, York H. Dobyns, and G. Johnston Bradish. "Correlations of random binary sequences with pre-stated operator intention: A review of a 12-year program." *Journal of Scientific Exploration, 11,* No. 3 (1997). pp. 345–367.

[7] Brenda J. Dunne, York H. Dobyns, Robert G. Jahn, and Roger D. Nelson. "Series position effects in random event generator experiments." *Journal of Scientific Exploration, 8,* No. 2 (1994). pp. 197–215.

4

REG EXCURSIONS

I love fools' experiments, I am always making them.
— Charles Darwin[1]

The Benchmark REG experiments described in the previous chapters constituted the most extensive database achieved at PEAR over its 30-year history, but they were substantially complemented by many other investigations of anomalous human/machine interactions. Some of these were pilot studies searching for new candidates for more intensive and detailed research expeditions; some addressed specific issues of strategy, technology, or protocol; some attempted to identify more propitious feedback modalities; some probed for possible pragmatic applications; some were successful; some were not. Although several of these employed categorically different types of random physical sources, all of those described in this chapter utilized the same basic microelectronic REG technology as in the Benchmark studies, or a close variant thereof.

A. Co-Operator Experiments

Niels Bohr's celebrated Complementarity Principle maintains that atomic-scale events can be described in two different frames of reference—as time- and space-localized particles, or as diffuse "probability-of-observation" waves. In our own theoretical musings we invoked this and other quantum mechanical concepts as metaphors to represent the subjective dimensions of human experience in an attempt to clarify some of the objective physical phenomena commonly regarded as "anomalous."[2] Specifically, we explored the possibility that our tendency to represent these in particulate languages and coordinate systems might obscure some of their essential wave-like qualities that may underlie the anomalies. To test this hypothesis, we performed a body of REG experiments that required two experienced operators, termed "Co-Operators," to perform the task in concert, in an attempt to discern whether the combined results displayed particle-like linear combinations of their individual efforts, or indications of wave-like superpositions with their characteristic diffraction, interference, and barrier penetration capabilities.[3]

These co-operator experiments comprised a total of 85,000 trials per intention generated by 15 operator pairs in 42 independent series. Briefly summarized, the results yielded an overall HI − LO separation in the intended directions that was modestly significant ($z = 1.883$, $p < .03$), but none of the cumulative deviation patterns of specific operator pairs showed indications of simple linear combinations of the individual signatures. In fact, some individually successful operators produced null effects when working as co-operators, while others with minimal or even negative individual results produced strong positive yields as pairs. Yet, the signatures produced by repeated efforts of given co-operator pairs tended to be relatively consistent, suggesting that the superpositions were more wave-like than particulate in character.

This study also provided unexpected indications that the gender pairings of the individual operators might be important factors contributing to their co-operator performances. For example, as shown in Figure II-10, operator pairs of the same sex tended to produce null results, actually trending insignificantly in the directions opposite to intention. Opposite-sex pairs, on the other hand, tended to produce significant overall results in the desired directions, with effects approximately twice as large as those generated by these same individuals working alone. Most strikingly, this enhancement of effect size appeared strongest when the two operators shared a deep emotional bond with each other. In those cases, the results were nearly seven times larger than those produced by the individual efforts. A subsidiary feature of the opposite-sex data was a more balanced correlation between high and low achievements, compared to the asymmetrical yields frequently observed in the single-operator experiments, especially for female operators, wherein one direction of intention often produced considerably stronger results than the other.

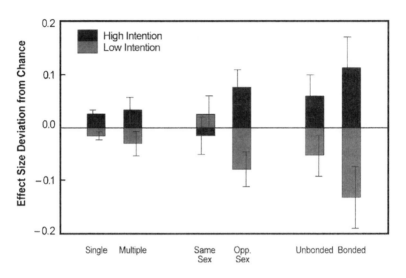

Fig. II-10. Block graph display of effect sizes for various categories of co-operator results (1-σ error bars superimposed).

Such results injected one or more of three possible implications into our search for correlates and phenomenological understanding: 1) collective effects do not compound linearly; 2) interpersonal resonance is an important subjective factor in establishing a propitious environment for anomalous human/machine interactions; 3) operator gender, *per se*, may be an important correlate of the phenomena.

B. PEAR 200: An Operator Competition

The results of the Co-Operator experiments raised the logical question of what would happen if two operators were in a competitive, rather than a cooperative mode. An exploratory protocol was designed, termed "The PEAR 200," in which the output from the REG was divided into two separate data streams, with a feedback display in the form of a rudimentary "racetrack" consisting of two "tracks" of contrasting colors, each assigned to one of the streams. A moving "car" traveled around each track, driven by the output data in such a way that when the operator assignment was HI, the accumulation of high counts would propel the respective car and low counts would slow it down, and vice versa for the LO assignment. The protocol for a given series consisted of two runs of 100 trials each in the HI and LO assignments, with both operators following the same directional assignment in each run.

PEAR 200 feedback display

A total of 89 series, or 356 races, were generated for a composite database of 35,600 trials per intention. Of these, 20 male operators participated in 130 races and 9 females in 188 races, each in competition with another human operator. In addition, some 38 races were run by individual operators striving against a random REG output stream displayed on the computer monitor, without a human opponent. For the composite human/human competitions, both the male and female operators produced HI and LO results that were comparable with, or somewhat larger than, those observed in the Benchmark data. None of these effects were statistically significant, however, due to the modest size of the database. Competitions with the computer, on the other hand, yielded LO results in both directions of effort. Despite the ambiguous outcome of this experiment, there was little doubt that the operators found it highly enjoyable, if the simulated engine sounds coming from the experimental room during the races were any indication!

C. ArtREG

Another REG-based experiment introduced with the hope of engendering stronger resonance between the operators and the device, and thence larger anomalous effect sizes, utilized works of art as both the physical targets and the feedback modality. In this study, operators selected two pictures from a library of 24 scanned images, which were then superimposed on the computer screen with half the pixels initially assigned to each picture. Output from one of our portable microelectronic REGs[4] determined the subsequent proportions of assigned pixels while the operator attempted to influence one of the pictures to dominate the screen. Data were generated in series comprising two runs of 100 trials, each designated HI or LO, wherein trials achieving the assigned directions caused enhancement of the chosen image. More than 50 operators accumulated over 400,000 trials on this ArtREG experiment, and while most found their experiences enjoyable, the overall results showed little significant correlation with their pre-stated intentions.[5]

Once again, however, unanticipated secondary effects appeared in the result patterns for individual operators and particular pictures. For example, of the twenty-four available images, only one or two would be expected to produce significant yields simply by chance, but in fact six displayed extra-chance results. Moreover, those six images shared the common feature that all entailed some form of sacred imagery, while the others were of largely mundane content. The results from the religious subset actually compounded to effect sizes comparable with the

Benchmark REG data, although the number of trials they comprised was insufficient to establish statistical significance.

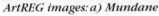
ArtREG images: a) Mundane   *b) Sacred*

An *ad hoc* student experiment performed subsequently to pursue this puzzling correlation presented operators with two competing images, one of which was an explicitly Christian work of art, set in competition with a computer graphic. The study yielded a composite z-score for the HI − LO separation in excess of 2.2 (p = .013) for the religious picture, which, given the small scale of this experiment, reflected an exceptionally large effect size. All of the significant performance in this experiment, however, was contributed by three of the operators whose religious persuasion was Christian; the two non-Christian operators scored only at chance.

One heavily utilized variant of the ArtREG protocol allowed the operators to select just one image from the library, which competed with a multi-colored random pixel illumination, with the effect that the chosen image appeared to be emerging from, or diffusing into, this random noise background. Although this data subset also displayed little overall statistical anomaly in the laboratory-based experiments, the concept proved attractive for major applications elsewhere, such as the Trapholt experiment mentioned in Section I, in which this technology and its theoretical premise was used to establish a communal experiment as an artistic installation at a prominent art museum. The publicly accessible display featured an engaging image of a baby, contrasting with a random background, and visitors were encouraged to attempt to sharpen the baby image, while the equipment recorded and compounded the accumulating digital data. Over a full year of active display, some 29 million trials (each the summation of 200 random bits) of active data were collected, interspersed with almost 3 million calibration trials generated with an unchanging neutral screen. As shown in Fig. II-11, the active data showed a marginally significant mean shift in the direction of greater image clarity with a z-score of 1.634 (p = .05). The local calibrations produced a nominal z-score of −0.255.

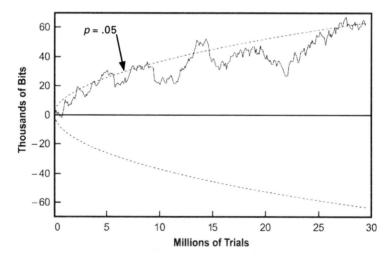

Fig. II-11. Trapholt experiment data summary: Deviation from chance mean vs. number of trials accumulated.

While the empirical yield of this experiment was only marginally supportive of the primary hypothesis, the higher goal of the artists to illustrate the intrinsic compatibility—indeed the intrinsic complementarity—of art and science, was deemed to have been well satisfied. The installation received high popular acclaim and subsequently was transported to a museum in Stockholm where it was on display for a year. It was later updated with more sophisticated technology and data processing capabilities in collaboration with scientists of the Niels Bohr Institute, for exhibit as an active experiment in the Esbjerg art museum in Denmark.

D. The Yantra Experiment

Several of the *ad hoc* experimental designs just described concurred on the evidence that direct conscious feedback, however attractive, did not necessarily enhance operator performance. Indeed, in some cases it appeared to impede it, a possibility consistent with some of the theoretical postulates proposed in Section IV regarding the role of unconsciousness processing. Another sequence of empirical studies, termed "Yantra," was therefore designed and implemented to test this hypothesis directly.[6,7]

In this experiment, no outcome-related feedback of any form was given to the operators, and no concurrent inspection of results by them was permitted. Rather, an assortment of more subtle visual and auditory environments were provided, which were hypothesized to be *subconsciously* conducive to anomalous performance. The choice of visual patterns to appear on the operators' computer screen was derived from the earlier experience with the ArtREG experiment, wherein it had been found that target pictures involving aspects of religious or spiritual imagery seemed to be disproportionately effective in that venue. Consistent with the caveat to minimize conscious processing and analytical focus, a mandala design know as "Sri Yantra" was utilized as the basic numinous display:

Yantra experiment: Sri Yantra pattern

In this familiar pattern, the core of interlocking triangles is intended to represent the interpenetration of spirit and the material world; and the abstract framing elements surrounding the core commonly appear in other mandala designs as well. Operators were offered three choices of display: 1) a static picture comprising white lines on a blue background; 2) a dynamical sequence of changing colors of the pattern segments that was driven by an auxiliary pseudorandom REG program; or 3) a blank computer screen. Several options

Yantra experiment: Operator feedback display

for a concomitant audio environment also were available: 1) regularly spaced drumbeats, synchronized with the mandala color changes; 2) double drumbeats, similarly synchronized, similar to a heartbeat; 3) operator-provided recorded music; or 4) silence.

Although the HI and LO intentions for the REG output were not correlated with the feedback display, the protocol maintained the volitional/instructed option offered in most other REG experiments. Thus there were 24 possible protocol combinations, which in this application were selected solely on operator preference, and hence compounded to a severely inhomogeneous data matrix. All told, over 60 operators generated more than 1000 formal series, each series comprising two runs of 100 trials under the HI intention, and two in the LO. (No baselines were collected, since only matched pairs of HI – LO separations were considered indicative in the analyses.)

The elaborate data analysis methods and their results are described in detail in Ref. 6. Although the mean shift results were largely unimpressive, the data patterns were replete with structural idiosyncrasies, and a number of chi-squared formats were invoked to examine and clarify them. The most general test for distinctive individual behavior utilized a chi-squared criterion constructed from the z-scores for every data segment for which the combination of operator, intention, and feedback were held constant. These yielded a probability against chance of $p = 0.020$, with the major portion of the anomaly driven by the high-intention segments [*cf.* Fig. II-12(a,b)]. Significant gender differences, similar to those found in many other REG studies, also were apparent, and the individual operator data encompassed an excess of extreme outliers at a chance likelihood of less than $p = 0.005$.

The severity of the structural disparities in this database can better be illustrated by comparing the probability of the feedback segment chi-squared values computed by standard statistical F-tests,[8] as displayed in Table II-2, which lists the probabilities thus derived vs. chance for the HI data alone, which was the more distorted of the two intentions:

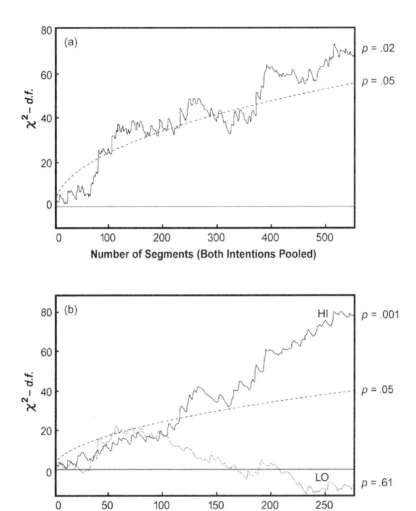

Fig. II-12. Cumulative plots of (χ^2 – d.f.) for Yantra experiment:
a) Distinguished by operator, environment, and intention;
b) Distinguished for operator × environment segments, HI and LO
separations.[6,7]

Note the wide disparities of results over the matrix of feedback options chosen by the operators, and that many of these segments,
including some of the most popular choices, produced little or no

anomalous yield. It again follows that the aesthetic preference for particular pattern combinations was no guarantee of their facilitation of anomalous performance, a lack of correlation that was consistent with the PEAR200 and ArtREG experiences.

One final pertinent note concerns the inclusion of this particular experiment in the agenda of an *ad hoc* sequence of efforts by three visiting

TABLE II-2

Chance Probabilities of Chi-Squared F-Test Values for Yantra Experiment Feedback and Protocol Options (HI Data Only)[7,8]

Assignment	# Series	χ^2 Probability
Instructed	198	0.0005
Volitional	81	0.30
Visual Environment		
Static	77	0.0001
Changing	157	0.323
None	45	0.057
Audio Overlay		
Regular beat	100	0.0044
Heartbeat	84	0.026
Other	15	0.071
None	80	0.43

practitioners of a Japanese healing discipline known as Johrei. As reported in detail in a dedicated technical note,[9] these operators employed their Johrei techniques for one half of their experiments, and refrained from using them in the other half. All of their non-Johrei efforts compounded well within chance, but when using their Johrei strategies, coupled with the instructed, static image, single-beat protocol option, two of the three achieved anomalous effect sizes an order of magnitude larger than those produced by the body of common operators. Curiously, however, these dramatic results were *opposite* to their stated intentions ($z_\Delta = -3.53$; $p_\Delta = .0004$). (The third operator, who did not opt for this particular protocol option, achieved only chance results in both his Johrei and non-Johrei efforts.) While this Johrei subset of the Yantra database is far too small to support any systematic correlations of this healing technique with the secondary and structural parameters of the data distributions, it does vigorously underscore the idiosyncratic character of the plethora of anomalous effects that have continued to manifest throughout the battery of specific experiments described in this chapter, and to some extent recoverable from the prior Benchmark studies.

In searching for conceptual metaphors for this behavior, one might offer the impact shattering of a brittle solid, or the unstable transition of a smooth laminar fluid flow into a complexly chaotic turbulent stream, in each of which essentially identical initial physical conditions are known to spawn a myriad of unpredictable subsequent states and systemic configurations. For these situations, the relevance of contemporary chaos, complexity, and other non-linear dynamical theories becomes suggestive, the conceptual utility of which may prove stimulating in attempts to model the empirical behavior.

E. MegaREG

Most of the PEAR REG data, both Benchmark and beyond, suggested that the anomalous effects could most parsimoniously be regarded as deriving from a direct alteration of the probabilities of the elementary binary events presented in the particular experiments, rather than from more complex dynamical mind/matter interactions.[10,11] If this is true, it then follows from elementary statistical theory that the anomalous yield of any given experiment should increase proportionally to the square root of the total number of bits processed, all other factors held constant. The early Benchmark experiments had included two variants of their usual 200 bit/trial format (REG-200),* extending them to 20 bits/trial (REG-20) and 2,000 bits/trial (REG-2000), respectively, all drawn from the same noise source but with the sampling rates changed to keep the spacing of individual trials roughly the same (~1/sec).[12] Although the amount of data acquired in these REG-20 and REG-2000 versions was considerably smaller than that in the Benchmark REG-200, the scales of mean shifts, summarized in Table II-3 along with the REG-200 results for comparison, seemed roughly compatible with the bitwise effect hypothesis, although the LO data in the REG-20 subset, while well beyond chance, were curiously opposite to intention.

---

*REG-200 should not be confused with the PEAR 200 data.

TABLE II-3

Results of Early 20- and 2000-Sample REG Explorations

	# Trials	Mean	Std. Dev.	z-Score	Probability
REG-200 Data					
High	839,800	100.026	7.070	3.369	3.77×10^{-4}
Low	836,650	99.984	7.069	−2.016	.0219
Baseline	820,750	100.013	7.074	1.713	.0867
HI − LO Diff.	—	—	—	3.809	7×10^{-4}
REG-20 Data					
High	41,750	10.012	2.235	1.097	.136
Low	40,250	(10.028)	2.254	(2.481)	(.007)
Baseline	41,750	10.007	2.225	0.610	.271
HI − LO Diff.	—	—	—	(−0.956)	.830
REG-2000 Data					
High	80,100	1000.155	22.342	1.956	.025
Low	82,200	999.853	22.300	−1.889	.029
Baseline	75,500	1000.018	22.245	0.227	.410
HI − LO Diff.	—	—	—	2.718	.003

Note: Numbers in parentheses ( ) indicate effects opposite to intention.

While these results were intriguing, the relatively small scales of the REG-20 and REG-2000 databases and the limited technical and personal resources available to us at that time restricted our ability to extend these studies to establish the generality of the observed effects or to define their range of applicability. Several years later, as our information processing technology matured, it became possible to mount a more aggressive empirical assessment of this question using a major modification of the REG equipment that permitted the trial size to be increased from the usual 200 bits to 2,000,000 bits.[13-16] This factor of 10^4 increase in this model thus could have been expected to enhance the statistical yield 100-fold for a comparable number of trials.

The user interface for the data-collection program was unaltered from the earlier REG experiment in order to ensure that the subjective experience of the operators was changed as little as possible. As before, trials were collected and presented to the operator at a pace of about one per second, and for any given experimental series, the operator had the option of a numerical trial value feedback, a graphical cumulative-deviation trace feedback, or no feedback at all. And to obviate possible psychological response differences to the larger experimental values (mean 1,000,000; standard deviation 707.1), the device was designed randomly to intersperse the 2,000,000-bit trials with 200-bit trials, both drawing from the same noise source. In each format, 2,000,000 bits were collected and processed, but in the first case all the bits were summed, and in the other only every ten-thousandth bit was used for the sum. The expected mean and standard deviation of the former were arithmetically manipulated to return them to the values of the latter (100 and 7.071, respectively) in order to maintain the perceived indistinguishability of the two trial types. In all other respects the usual protocol was followed: each series comprised one thousand trials in each of the HI, LO, BL conditions, and trials were collected in runs of continuous data generation of length selected by the operator, *i.e.* 50, 100, or 1,000 trials per run.

As detailed in the references, the calibration output characteristics of this "high density" equipment departed slightly from idealized theoretical binomial combinatorials, requiring empirical null determinations, the corresponding use of statistical *t*-scores rather than *z*-scores, and the definition of generalized independent measures of differential HI − LO mean shifts.[14,15] Short of recounting these analytical details in full, the bottom-line results of the simulated 200-bit trial data were now indistinguishable from chance, while the 2,000,000-bit data produced a *t*-score of −3.90, with an associated probability of 9.4×10^{-5} (2-tailed), but in the direction *opposite* to intention. In short, we were faced with experimental data that were grossly inconsistent with the bitwise probability hypothesis, but also totally failed to replicate the huge pool of prior 200 bit/trial Benchmark results, to the extreme of showing no significant HI − LO separation at all!

All of this bemusement predicated a further attempt at empirical replication we labeled "MegaMega," which used the same MegaREG equipment but processed only the fully sampled 2,000,000 bit/trial outputs (hence it returned no 200 bit/trial simulations for comparison). These MegaMega results conspired to deepen the mystery: there now appeared a significant performance asymmetry between all of the intentional data and their proximate baseline results. As in many of our other experiments, this asymmetry was totally attributable to the female operators.[17] A host of subsidiary structural and individual operator analyses revealed that these distorted results were broadly established across the operator populations. No other significant secondary correlations emerged to enlighten the mystery.

If the MegaMega and MegaREG data were combined, their HI − LO t-score was lowered to −4.03, and the corresponding 2-tailed chance probability to 5.65×10^{-5}. Sparing the reader much of the detailed calculations, the absolute value of the corresponding 2,000,000 bit/trial effect size for the pooled data was some −2.77 times as large as the Benchmark value, which in turn transcribed to an effect size per bit of −0.277 times that of the original Benchmark data, but *opposite* to intention! As a consequence, we could not avoid the conclusions that raising the bit count per trial from 200 to 2,000,000, in complete contradiction to the bitwise hypothesis, appeared to have produced: a) an inversion of the sign of the effect, b) only an approximately three-fold increase in statistical yield per trial, and c) a roughly 30-fold decrease in statistical yield per bit, with all three relations well established empirically.

The theoretical implications of these MegaREG explorations thus impelled us yet again toward concession of incomplete understanding of all the pertinent factors bearing on the phenomena or, yet more drastically, toward some vast generalization or even abandonment of our sacred scientific causal determinism.

F. Variance Variation

The primary variable in the Benchmark experiments, and indeed in virtually all of the subsequent human/machine experiments, was the hypothesized shift of the output distribution mean in the direction of pre-stated operator intention. However, as later studies began to reveal unusual distortions in higher moments of the distributions and correlations with certain secondary parameters of the protocols, the question of whether these mean shifts were an intrinsic property of the anomaly itself, or whether other distortions also might be specifically coupled to the experimenters' and/or operators' particular intentions or expectations. In addition, our investigations of gender differences, as described in Chapter II-13, indicated that in the mean-shift efforts, females tended to display larger variances and more asymmetrical results than males, and we were curious to see whether these tendencies would persist when operator intention was addressed more directly to the distribution variance.

This question was explored in two modest experimental excursions in which operators attempted to alter the variances of the REG output distributions, rather than their means. That is, instead of trying to obtain a succession of counts higher or lower than the theoretical mean value of 100, they endeavored to produce more extreme or more constricted count values, regardless of their direction, thereby increasing or decreasing the standard deviation of the trial score distribution from its theoretical value of 7.071. A successful "HI" effort would thus entail production of a wider range of count values, while a "LO" intention would attempt to keep the values closer to 100.

In the first experiment, two operators, both female, generated a total of approximately 40,000 trials per intention using the Benchmark REG device. These yielded small, but non-significant trends in the desired directions, *e.g.* standard deviations of 7.083 in the HI (p = .321) and 7.059 in the LO (p = .320) intentions. A second experiment carried out by 16 operators (11 females and 5 males) comprised a total of 326 50-trial runs, or 8,150 trials per intention. Of these, 186 (57%) of the LO-intention runs produced standard deviations that were smaller than the theoretical expectation of 7.071, yielding a z-score of 2.547 (p = .007), but only 47% of the HI-intention runs produced results in the desired direction, corresponding to a non-significant negative z-score of –1.114. Secondary analyses of the gender disparities indicated that once again the females were primarily responsible for the asymmetry and for the apparent success of the LO intention results.

Given the small size of these databases and the unbalanced number of male and female participants, these results also were far from conclusive and their implications were unclear, but some aspects may have been indicative. For example, as in the mean-shift experiments, the females continued to affect the output variances more than the males, and their results continued to be strongly asymmetrical, with 58% of the LO runs vs. only 46% of the HI consistent with intention. Perhaps more importantly, there appeared to be no notable correlations between the variance effects and the distribution means. In short, these explorations indicated that the character of the anomalous results were more closely coupled to the focus of the operator's intention, rather than being constrained to some inherent mean shift propensity of the anomaly itself.

G. Benchmark Redux

The growing blizzard of idiosyncratic structural anomalies that emerged in several of the later REG experiments in place of the cleaner primary HI – LO data separations of the Benchmark studies predicated an empirical return to the possibility of device dependence of these human/machine effects. It was noted that whereas most of the strong earlier results had been achieved on the original large REG Benchmark machine, where digital feedback had been presented on the face of the device, for reasons of operational expediency most of the later experiments had utilized smaller and simpler units, such as Johnson noise sources[18] and miniaturized microREGs, with feedback provided by computer transcription of the data to a visual display. It seemed reasonable to speculate whether a return to the original equipment and pro-tocols would restore the original results, thereby implicating the simpler devices as causes of the replication problem.

To permit as rapid an assessment of this issue as possible, a pre-specified body of new experiments was posted, wherein the original REG source was deployed in a protocol closely emulating that of the earlier Benchmark subset we had called "ThouREG." In this exploration, labeled "OREG" (implying "old" or "original" REG), 200-bit trials were accumulated in 1,000-trial series, all performed in single runs. The yield of this new study could then readily be compared with that of its ThouREG predecessor.

As summarized in Table II-4, the primary HI – LO OREG re-sults in fact were completely consistent with chance, and there-fore totally incompatible with the prior ThouREG data, resound-ingly implying that the demise of the principle HI – LO anomaly could not be attributed to the succession of equipment changes, *per se*, but again raising the chimera of the failure to replicate the original 200-bit results. Once more, however, a host of structural distortions were observed to emerge in replacement of the gross HI – LO separations: chi-squared issues (both goodness-of-fit and variance aspects) abounded; gender disparities were rampant, but not consistent with previous evidence; and various auto-correla-tions, not expressed in the baseline and calibration data, raised further suspicions of pertinent subjective factors.

Table II-4

Z-Score Comparisons (p_z) for Benchmark, ThouREG, and OREG

	All Benchmark	ThouREG	OREG
HI	$3.37 (4 \times 10^{-4})$	2.38 (.009)	−0.34 (n.s.)
LO	−2.02 (.022)	−1.67 (.048)	−0.47 (n.s.)
HI – LO	$3.81 (7 \times 10^{-5})$	2.91 (.002)	0.09 (n.s.)

H. ProbREG

Most of our applications of REG equipment to human/machine anomalies experimentation proceeded under two presumptions, one technical and one phenomenological, *i.e.*: 1) the intrinsic binary probability of the individual digits or bits prepared and presented by the machines in their unattended (calibration) operations was precisely p_b = 0.5, and 2) any anomalous effects in the machine outputs achieved by the operators could be represented as relatively small changes in those binary probabilities, Δp_b, which, while clearly operator specific, remained constant or changed only slightly over long sequences of a given operator's random output data streams. Convenient as these assumptions may be for experimental design and data analysis, they are not necessarily sacrosanct, and empirical and analytical challenges to them may potentially enable more effective experimental interactions, and better illuminate the basic nature of the phenomena. Two attempts were mounted to probe such assumptions, and to extract from them some deeper understanding of the dynamics of these mind/matter interactions.

For study of the first category of parametric sensitivity, we designed, constructed, analyzed, and utilized a variant of our microREG devices termed "ProbREG," which allowed three separate initial binary probability settings of p_b = .0625, .5000, and .9375, respectively.[19,20] As detailed in the references, this equipment was then deployed in a proof-of-concept experiment that required each of five experienced operators to generate datasets of 2500 × 200-bit trials under pre-stated HI, LO, and BL intentions for each of the three binary settings. The data were then compared analytically with various hypotheses for the dependence of Δp_b on its parametric value setting, p_b, *e.g.*, $\Delta p_b \propto p_b$; $\Delta p_b \propto 1/p_b$; Δp_b = constant; etc.

As encountered in several other *ad hoc* studies, the empirical results once again conspired to confound what had seemed to be a well-defined theoretical anticipation. Despite the close similarity of this REG equipment to our standard microREG technology, feedback modalities, and operator protocols, and the close conformity of more than 12,000,000 calibration trials to theoretical chance expectations, the active data displayed much smaller anomalous mean shifts than those found in our earlier Benchmark studies, *even for the p_b = 0.5 setting of the new equipment.* This inexplicable bashfulness of the basic REG anomaly to re-manifest in this context not only largely disabled the desired comparisons with the various Δp_b vs. p_b models, it comprised yet another example of the phenomenological elusiveness encountered in several other databases. And as in other cases, it once again offered in its place a number of secondary structural features that were quite anomalous in their own rights.

Most notable of these new aberrations was a drastic lack of independence among the terminal z-scores of the intention/operator/p_b data matrix, which tended to cluster excessively around the chance values to an extent approaching, and in some cases exceeding, statistical significance. The scale and structure of this deviation was confirmed by an extensive Monte Carlo simulation based on a sequence of iterations of the calibration data, which in their adherence to normally distributed chance behavior excluded any equipment malfunction as a primary source of this sub-anomaly. This in turn posed a very difficult phenomenological/psychological issue. Namely, by what conceivable means could five different operators, endeavoring to utilize their individual techniques to achieve anomalous HI – LO mean-shift separations from an experimental target configured to three widely disparate probabilities, unconsciously conspire to produce results that were substantially more correlated among themselves than should be expected by chance, even when they compounded to statistically negligible primary effects? Bizarre as such secondary anomalies may appear, as noted earlier we have encountered many forms of them in various other experimental contexts, leading us to suspect that such are legitimate alternative expressions of the operator-induced anomalous behavior that will need to be accommodated in any comprehensive model of the phenomena.

All of this mystery notwithstanding, two other more pragmatic insights were derived from the study. First, despite the small effect sizes, it became clear that the better criterion for comparison of the effectiveness of various experiments of this particular class was their standard statistical z-scores, rather than the raw mean shifts, $\Delta\mu$, or even the non-dimensionalized normalizations thereof, $\Delta\mu/\mu$, *per se*. As derived in the references, this in turn implied that the Δp_b values attainable by the operators should scale with a more generalized bit-probability, defined as $\mathcal{P} = \sqrt{p_b - p_b^2}$, rather than simply with p_b, a result which implicitly introduced a "noise-to-signal" ratio of the data stream, σ/μ, as the pertinent parameter governing the interaction.

Finally, from an operational perspective, we reluctantly were forced to concede that neither of the extreme initial p_b settings, .0625 or .9375, induced sufficiently radical departures from the $p_b = .5000$ ProbREG data to encourage the refinement of the device and the collection of the order-of-magnitude larger databases that would be needed to discriminate the original alternative hypotheses and structural features more authoritatively, let alone to attempt to exploit any $p_b \neq .5$ technologies for other research and application purposes.

I. Run-Length Dependence

A similar gambit attempted to address the possibility that the intrinsic operator-imposed binary probability shifts, Δp_b, individually distinctive as they might be, were subject to some characteristic form of gradual evolution over the course of extended data accumulations, beyond the shorter-term serial position effects described earlier. Here it was deemed unnecessary to construct any new equipment, but simply to examine retrospectively our very large established REG databases. Technically detailed reports of this analysis exercise are available in References 21 and 22, which include numerous graphical representations of experimental Δp_b evolutions over compounding database sizes (N), compared with Δp_b = constant and $\Delta p_b \propto 1/\sqrt{N}$ models, where the latter reflects the "noise-to-signal" hypothesis proposed in the ProbREG context. Once again, however, it quickly became evident that despite the huge databases available for these comparisons, there still was insufficient statistical power to distinguish convincingly between those alternative behaviors.

In the hope that this insensitivity was derived from the dilution of the composite databases by the preponderance of less effective participants, the analysis was repeated for two of our most accomplished prolific operators, but here again, no statistically robust distinction between the two models could be established. Alternatively, the composite databases were artificially homogenized into subsets of equal 50-trial segments allowing more sophisticated least-squares and chi-squared comparisons between them and the two fundamental models, but even these failed to yield adequate discrimination.

Curiously akin to the unsolicited appearance in the ProbREG results of a compaction in terminal z-scores, however, was the emergence from these run-length data of a similar constriction in the composite HI, LO, and HI − LO results, all of which displayed somewhat less variation among them than expected by chance for either of the models, yet was not evident in the baselines or calibrations. While the references (21, 22) indulge in some speculation about the implication of this substructure, it can only be added to the bemusing arsenal of structural anomalies these experiments have been observed to display.

J. Miscellaneous Excursions

Beyond the explorations just described, in due course we also undertook an assortment of other variations of the standard REG studies that turned out not to reveal any further insights than we had already found in previous studies, or for various technical reasons could not be pursued beyond initial pilot phases. Briefly summarized, these included:

1) a 1,000-trial per intention protocol termed "SigREG," which treated each operator's data as a distinct experiment, rather than pooling the results as we had done in other experiments;

2) a shortened version of the earlier protocol, requiring only two 100-trial runs per intention and taking approximately 20 minutes for a series, which we referred to as ShortREG;

3) a machine/machine design wherein two REGs were allowed to run concurrently overnight with no operator involved, to determine whether any unusual correlations might appear in their respective outputs (none were observed);

4) a multiple REG experiment where an operator attempted to influence one REG while a second device next to it ran passively with no display (again no correlations were noted in the outputs of the two devices); and

5) an exploration referred to somewhat whimsically as "VeggieREG," in which a large philodendron plant was placed in a dimly lit room with an REG connected to an overhead lamp, so that an increase in the cumulative mean output of the REG would result in increasing the available illumination. (In the first two days of operation this experiment produced a significantly high correlation; unfortunately, the next day the accompanying computer malfunctioned and could not be repaired, and the experiment was not pursued further.)

References

[1] Charles Robert Darwin. As quoted in E. Ray Lankester, contrib., "Charles Robert Darwin," in Charles Dudley Warner, ed., *Library of the World's Best Literature: Ancient and Modern.* New York: J. A. Hill & Company, 1902. p. 4391.

[2] Robert G. Jahn and Brenda J. Dunne. "On the quantum mechanics of consciousness, with application to anomalous phenomena." *Foundations of Physics, 16,* No. 8 (1986). pp. 721–772.

[3] Brenda J. Dunne. "Co-Operator Experiments with an REG Device." Technical Note PEAR 91005. Princeton Engineering Anomalies Research, Princeton University, School of Engineering/Applied Science, Princeton, NJ. December 1991.

[4] Roger D. Nelson, G. Johnston Bradish, and York H. Dobyns. "The Portable PEAR REG: Hardware and Software Documentation." Internal Document #92-1. Princeton Engineering Anomalies Research, Princeton University, School of Engineering/Applied Science, Princeton, NJ. 1992.

[5] Robert G. Jahn, Brenda J. Dunne, York H. Dobyns, Roger D. Nelson, and G. Johnston Bradish. "ArtREG: A random event experiment utilizing picture-preference feedback." *Journal of Scientific Exploration, 14,* No. 3 (2000). pp. 383–409.

[6] York H. Dobyns, John V. Valentino, Brenda J. Dunne, and Robert G. Jahn. "The Yantra Experiment." Technical Note PEAR 2006.04. Princeton Engineering Anomalies Research, Princeton University, School of Engineering/Applied Science, Princeton, NJ. October 2006.

[7] York H. Dobyns, John C. Valentino, Brenda J. Dunne, and Robert G. Jahn. "The Yantra experiment." *Journal of Scientific Exploration, 21,* No. 2 (2007). pp. 261–279.

[8] George E. P. Box, William Gordon Hunter, and J. Stuart Hunter. *Statistics for Experimenters: An Introduction to Design, Data Analysis, and Model Building.* Wiley Series in Probability and Mathematical Statistics. New York: John Wiley & Sons, 1978.

[9] Robert G. Jahn, Brenda J. Dunne, and York H. Dobyns. "Exploring the Possible Effects of Johrei Techniques on the Behavior of Random Physical Systems." Technical Note 2006.01. Princeton Engineering Anomalies Research, Princeton University, School of Engineering/Applied Science. Princeton, NJ. January 2006 (30 pages).

[10] York H. Dobyns. "Overview of several theoretical models on PEAR data." *Journal of Scientific Exploration, 14,* No. 2 (2000). pp. 163–194.

[11] Robert G. Jahn, York H. Dobyns, and Brenda J. Dunne. "Count population profiles in engineering anomalies experiments." *Journal of Scientific Exploration, 5,* No. 2 (1991). pp. 205–232.

[12] Robert G. Jahn. Brenda J. Dunne, and Roger D. Nelson. "Engineering anomalies research." *Journal of Scientific Exploration, 1,* No. 1 (1987). pp. 21–50.

[13] Michael Ibison. "Evidence That Anomalous Statistical Influence Depends on the Details of the Random Process." Technical Note PEAR 97007. Princeton Engineering Anomalies Research, Princeton University, School of Engineering/Applied Science, Princeton, NJ. October 1997.

[14] Michael Ibison. "Evidence that anomalous statistical influence depends on the details of the random process." *Journal of Scientific Exploration, 12,* No. 3 (1998). pp. 407–423.

[15] York H. Dobyns, Brenda J. Dunne, Robert G. Jahn, and Roger D. Nelson. "The MegaREG Experiment: Replication and Interpretation (revised edition)." Technical Note PEAR 2002.03. Princeton Engineering Anomalies Research, Princeton University, School of Engineering/Applied Science, Princeton, NJ. October 2002.

[16] York H. Dobyns, Brenda J. Dunne, Robert G. Jahn, and Roger D. Nelson. "The MegaREG experiment: Replication and interpretation." *Journal of Scientific Exploration, 18,* No. 3 (2004). pp. 369–397.

[17] Brenda J. Dunne. "Gender differences in human/machine anomalies." *Journal of Scientific Exploration, 12,* No. 1 (1998). pp. 3–55.

[18] Frederick Reif. *Fundamentals of Statistical and Thermal Physics.* New York: McGraw-Hill, 1965. pp. 589–594.

[19] Robert G. Jahn and John C. Valentino. "Dependence of Anomalous REG Performance on Elemental Binary Probability." Technical Note 2006.02. Princeton Engineering Anomalies Research, Princeton University, School of Engineering/Applied Science, Princeton, NJ. August 2006.

[20] Robert G. Jahn and John C. Valentino. "Dependence of anomalous REG performance on elemental binary probability." *Journal of Scientific Exploration, 21,* No. 3 (2007). pp. 473–500.

[21] Robert G. Jahn and York H. Dobyns. "Dependence of Anomalous REG Performance on Run Length." Technical Note 2003.01. Princeton Engineering Anomalies Research, Princeton University, School of Engineering/Applied Science, Princeton, NJ. February 2003.

[22] Robert G. Jahn and York H. Dobyns. "Dependence of anomalous REG performance on run length." *Journal of Scientific Exploration*, *21*, No. 3 (2007). pp. 449–472.

5

FIELDREG:
RANDOMNESS AND RESONANCE

For whenever two or more of you are gathered in My name
There is Love, there is Love.
— Paul Stookey[1]

For most of the studies described in the preceding chapters the basic hypothesis underlying the protocol designs and statistical analyses was that human intention, *e.g.* need, desire, volition, etc., could influence the output of a random physical process. Although the empirical results did indeed establish significant correlations with pre-stated operator intentions, as we acquired more empirical experience and began to explore theoretical models that might help explicate such anomalous effects, it became evident that correlates other than intention could be propitious for manifestation of the phenomena. For example, throughout this experimental era numerous informal operator reports alluded to a requisite sense of "resonance" between the operator and the device, task, or purpose, whenever anomalous results were being achieved. This is certainly consistent with the prevailing regard for such an ingredient in many forms of creative scholarly, artistic, spiritual, and interpersonal transcendence.

A. FieldREG Forays
While it had been relatively straightforward to design a protocol that had as its primary variable a pre-recorded intention to produce an effect in a stated direction of effort, developing a strategy that could control for subjective resonance was a more challenging task. Changing one's conscious intention is a relatively uncomplicated cognitive effort, but the establishment of an emotional

bond is a far more complex process that takes place at a deeper level of experience and does not lend itself so much to deliberate manipulation as to ego-effacing investment. Hence it seemed better first to identify a resonant operator situation and then to introduce a technological detector into that environment. To do this, we developed a simpler, portable REG device and a substantially different REG protocol that permitted passive monitoring of selected environmental situations and sites where human consciousness, or some subtler passive features of the sites themselves, might conceivably be altering or organizing a surrounding "field" of potential information, most broadly construed, without specific intention or direct attention to the experimental devices by any human participants. This new experiment was dubbed "FieldREG," a double entendre that connotes both the deployment of these devices in "field" situations, and their ability to monitor changes in prevailing consciousness "fields."[2,3]

Such FieldREG systems, comprising a smaller-version portable REG and a palmtop computer with appropriate software, were first deployed as passive background monitors in a variety of group assemblies where unusual collective dynamics might be

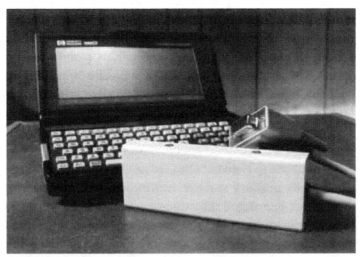

Portable FieldREG interfaced with palmtop computer

expected to ensue. In these applications, unobserved data were generated in continuous segments, with time-stamped indices identifying scheduled or unscheduled periods of particular interest. These were later examined for evidence of unusual behavior, as indicated by palpable shifts in the output means or by protracted periods of steady deviation. The behavior of the system in extended calibration or during extra-session intervals provided a form of on-line control data. Anomalous deviations were hypothesized to be indicative of some change in the prevailing information environment associated with the collective consciousness of the assembled group, or with the site itself.

One of our earliest FieldREG applications took place during an Academy of Consciousness Studies, a ten-day multi-disciplinary workshop for some 50 scholars involved in consciousness studies, which was held on the Princeton campus during the summer of 1994.[4] The FieldREG device was allowed to run in a continuous unattended mode throughout the various presentations and discussions. Figure II-13a shows the trace from one full day's operation (approximately nine hours) and includes the most extreme deviation associated with any session in the ten-day record. This segment, expanded in Figure II-13b, occurred during a discussion that addressed the pervasive presence of ritual in human activities ranging from everyday habits, to religion, to science, and was spontaneously described by the participants as being deeply engaging. The intrinsic chi-squared p-value for this segment was 0.00005, yielding a Bonferroni corrected value of 0.0028. (The Bonferroni adjustment provides a statistical correction for estimating the likelihood of a particular segment out of an assortment of multiple segments as evidence of an anomalous effect.)

Many similar deployments were subsequently attempted over several years of applications in spiritual or secular ceremonies and rituals, individual or group therapy sessions, business meetings, sporting events, musical and dramatic theatre, professional conferences, or any other group convocations that might include periods of unusually cohesive cognitive interaction, creative enthusiasm,

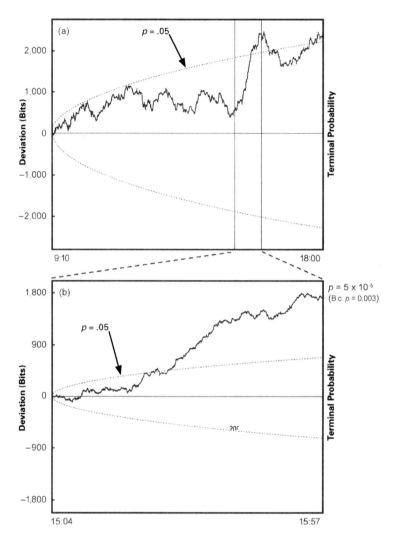

Fig. II-13. FieldREG response to an "Academy of Consciousness Studies": a) Full day's session; b) Marked 53-minute segment of (a) (*cf.* text).

or other forms of shared emotional intensity. One particularly striking example is a trace generated during a 20-minute healing ceremony performed by a Shoshone shaman at Devils Tower Monument in Wyoming.

Devils Tower, Wyoming

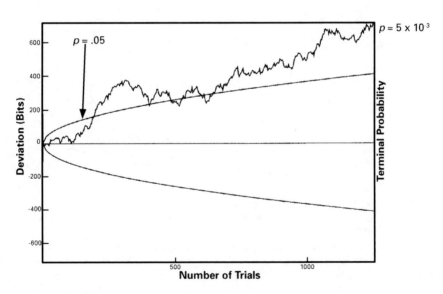

Fig. II-14. FieldREG response to a 20-minute healing ceremony performed by a Shoshone shaman at this site.

As yet it is unclear whether the direction of the deviations displayed in the traces is meaningful. Most applications show both positive and negative excursions, and our analysis explicitly ignores direction by considering only the variability (*i.e.* variance) of the deviations of the segment means.

Another productive FieldREG exploration, conducted over a two-week period, involved a group of 19 people who visited over 20 ancient Egyptian temples, pyramids, and sacred sites, and while there engaged in meditation and chanting in an attempt to create a spiritual connection with these places. As shown in Figure II-15, the composite yield was highly significant ($p = 4 \times 10^{-4}$). It may be interesting to note that one of these sites, the Temple of Philae, which had been moved from its original location during the construction of the Aswan Dam, produced a noticeably flat chance result.[5]

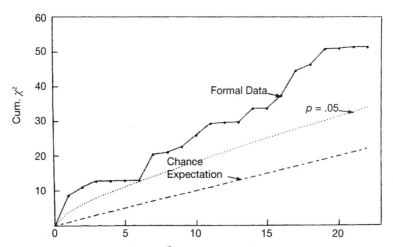

Fig II-15. Cumulative χ^2 results of FieldREG applications at 22 Egyptian sites.

B. The Role of Resonance

This possibility, the establishment of discernible "consciousness fields," or "information fields," has been widely proposed in many other contexts, by scholars from many disciplines,[6-9] and it seems consistent with the aforementioned testimony of our laboratory operators regarding their states of "resonance" with the experimental devices during successful sessions. In fact, hints of the possible importance of subconscious, collective, or environmental contributions to the anomalous phenomena, wherein explicit intention is relegated to a secondary or subtler role, have arisen in many other aspects of our empirical studies, including the ubiquitous series position effects (Chap. II–14), the observed deviations of non-intentional baseline data from chance behavior, an assortment of gender differences, the superior performance of some operators when not directly focusing on the machines, and the unusually large effect sizes of co-operator bonded-pairs.

A lengthy initial round of such FieldREG applications, which yielded many impressive segments of clearly anomalous responses, inspired an even more substantial replication phase[3] to search for some defining criterion for segregating the circumstances yielding strongly anomalous results from the remainder of the data.[2] The hypothesis posed to be tested was that trials performed in environments pre-specified to favor a high degree of subjective resonance within a participating human group would return larger effects or greater numbers of segments of anomalous response compared to those involving predictably less group resonance, *e.g.* captivating musical or dramatic performances, religious rituals, intense sporting events, or poignant public ceremonies, vis-à-vis boring lectures, mundane discussions, or meaningless formalities.

The results of this replication phase, like the earlier data processed retrospectively, were vigorously supportive of this hypothesis, to a remarkable degree. Figure II-16 shows the steady departure of a cumulative chi-squared data reduction[10] for the pre-specified resonant subset of all trials accumulated in both phases of the FieldREG experiments, compared with appropriate chance expectation and .05 chance probability reference envelopes. (The segregation of the two data phases is indicated by the dashed vertical divider.) These combined data concatenated to a chance

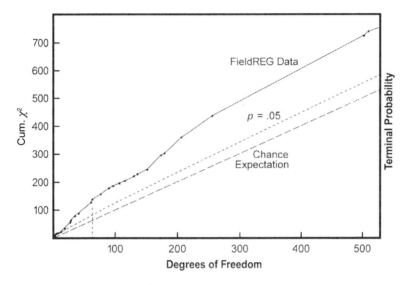

Fig. II-16. Cumulative χ^2 graph of FieldREG responses over the full set of applications predicted to be "resonant".

expectation of the order of 10^{-6}, and since the two phases did not differ significantly from each other, they constituted a strong *de facto* affirmation and replication of the hypothesis.

Equally, if not more remarkable, was the corresponding concatenation of the results of applications in the pre-designated "mundane" venues, where the chi-squared indicators compounded to a trace that lay well *below* chance expectation, to a significant statistical degree ($p = 0.019$; Fig. II-17), almost as if the devices had been anesthetized into tranquility by their immersion in these boring environments!

References 2 and 3 provide a broader catalogue of the observed character of FieldREG responses in a variety of application categories, including some for which the devices were not physically present at the designated sites. These "remote" applications again raised the generic issue of the relevance of objectively specifiable physical parameters in these phenomena. More specifically, the spatial and temporal independences seen in the

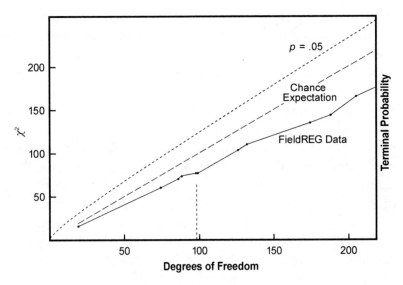

Fig. II-17. Cumulative χ^2 graph of FieldREG responses over the full set
of applications predicted to be "mundane".

laboratory-based REG experiments were found to carry over to
some of the FieldREG studies as well.

Any fully comprehensive search for more detailed criteria for
specification of effectively resonant venues of application will
be a formidable task, given the number and variety of potential
targets, the range and subjective character of the possible correlates to be considered, and the possible spatial and temporal protocol variants, and doubtless will require a prodigious database
management system to extract the salient correlations efficiently.
Notwithstanding, the inclusion of some form of resonance criterion in complementary status to that of intention seems inescapable, and will need to appear in any competent generic model of
these phenomena.

To recapitulate, on the basis of these somewhat preliminary,
yet complex observations, we were lead to postulate the existence
of a pervasive "consciousness information field" that may, under
certain circumstances, exhibit empirically detectable technical

modulations somehow stimulated by the participating individuals, groups, or sites themselves. In particular, it was suggested that via this field, human consciousness can act as a radiating source of information, capable of affecting otherwise random processes in the physical or psychical environment by inserting some degree of order that makes them slightly more predictable. Since the environmental aspects that seemed to correlate most strongly with such anomalous effects were largely subjective in character, this structuring influence, which might be labeled "subjective information," seemed to constitute an attribution of personal *meaning* to the specific situations or sites. In the field experiments reported here, as in the intention-based laboratory experiments, this modification of the consciousness information field appeared to manifest through alterations of statistical distributions generated by suitably instrumented physical systems that involved random or indeterminate components. Again, in the laboratory experiments, these alterations appeared to be correlated primarily with operator intention, wishing, or purpose, enhanced by some form of emotional or spiritual resonance. In the field experiments, resonance seemed to play the primary role, supplemented by some less conscious or less explicit form of intention.

The availability of newer and more versatile REG technology and operational software now makes it possible to extend FieldREG explorations into a broader range of potential venues. For example, one among many such applications is the Global Consciousness Project,[11] a PEAR derivative directed by Roger Nelson, which collects data continuously from a global network of REGs located in 65 host sites around the world, in a search for subtle correlations in their collective responses that may reflect the presence and activity of a broadly interconnective consciousness field.

In summary then, the FieldREG techniques offer promising vehicles for empirical assessment of natural situations wherever people are engaged in activities involving both their subjective and objective attention. Although by their nature such subjective properties are difficult to specify or monitor, let alone to quantify, we are persuaded that their inclusion is essential for fuller understanding of the anomalous interactions of consciousness with its environment. Interpretative challenges notwithstanding, the vision of a technology, however subtle and complex, that could reliably sense the degree of coherent purpose and productive resonance prevailing in such diverse human arenas as business and industry, healing and healthcare, public safety, education, athletics, and artistic performance, among countless others, and lead to beneficial applications therein, seems to justify extensive effort to bring it to fruition.

References

[1] Paul Stookey. *Wedding Song*. Public Domain Foundation, 1971.

[2] Roger D. Nelson, G. Johnston Bradish, York H. Dobyns, Brenda J. Dunne, and Robert G. Jahn. "FieldREG anomalies in group situations." *Journal of Scientific Exploration, 10*, No. 1 (1996). pp. 111–141.

[3] Roger D. Nelson, Robert G. Jahn, Brenda J. Dunne, York H. Dobyns, and G. Johnston Bradish. "FieldREG II: Consciousness field effects: Replications and explorations." *Journal of Scientific Exploration, 12*, No. 3 (1998). pp. 425–454.

[4] Brenda J. Dunne. "Report on the Academy of Consciousness Studies." With an Appendix by Robert G. Jahn. *Journal of Scientific Exploration, 9*, No. 3 (1995). pp. 393–403.

[5] Roger D. Nelson. "FieldREG Measurements in Egypt: Resonant Consciousness at Sacred Sites." Technical Note 97002. Princeton Engineering Anomalies Research, Princeton University, School of Engineering/Applied Science, Princeton, NJ. July 1997 (36 pages).

[6] Arthur Llewellyn Basham. *The Wonder That Was India: A Survey of the Culture of the Indian Sub-Continent before the Coming of the Muslims.* New York: Grove Press, 1959.

[7] Émile Durkheim. "Society and individual consciousness." In Talcott Parsons, Edward Shils, Kaspar D. Naegele, and Jesse R. Pitts, eds., *Theories of Society: Foundations of Modern Sociological Theory Vol. 2,* 720. Glencoe, Illinois: The Free Press, 1961.

[8] William James. *Human Immortality: Two Supposed Objections to the Doctrine.* Boston: Houghton-Mifflin, 1977. (Originally published 1898).

[9] Rupert Sheldrake. *A New Science of Life: The Hypothesis of Formative Causation.* Los Angeles, CA: J.P. Tarcher, 1981.

[10] George E. P. Box, William Gordon Hunter, and J. Stuart Hunter. *Statistics for Experimenters: An Introduction to Design, Data Analysis, and Model Building.* Wiley Series in Probability and Mathematical Statistics. New York: John Wiley & Sons, 1978.

[11] <http://noosphere.princeton.edu/>.

6

PSEUDORANDOM PARADOX

Anyone who considers arithmetic methods
of producing random digits is,
of course, in a state of sin.
— John Von Neumann[1]

One of the earliest questions addressed by the experimental program was whether the anomalies observed in the microelectronic REG context were restricted to particular sources of physical noise, or whether they could appear more ubiquitously across a broader variety of random physical processes. This issue was first explored by changing the electronic noise sources in the benchmark REG circuitry, and later by constructing a few categorically different types of physical random sources, all of which were submitted to the same basic differential (HI – LO) operator protocols and data analysis methods. Despite the broad range of character and scale of the physical equipment, the similarity of the results obtained on these devices suggested that the anomalous effects did not derive from physical influences on the noise sources, *per se*, but were more intrinsically manifested in the statistical *information* presented in their output data. This recognition thus posed the even more fundamental issue of whether such anomalous information re-ordering also could be achieved using *pseudorandom, i.e.,* deterministic sources.

Several attempts were made to address this question by replacing the physical microelectronic noise elements with circuitry that created binary signals that were technically deterministic in character, yet satisfied the same statistical criteria for randomness of their calibration distributions. The earliest effort utilized a feedback array of 31 microelectronic shift registers that produced a sequence of two billion bits cycling continuously with a repetition period of about 60 hours, so that the only non-deterministic aspect of the experiment should have been the time of incursion into the bit sequence initiated by the operator. Switched into the benchmark REG apparatus at an appropriate location, this source replaced the commercial noise diode and its immediate conditioning circuitry, but left all attendant sampling, counting, display and feedback circuitry, and software architecture identical to the physically random benchmark version. Thus, from the perspective of the operator, this system was indistinguishable from the Benchmark REG, and the experimental protocols employed were identical. The results of 29 series generated on this device indicated significant shifts in the means of the output distributions in the desired directions of effort (Fig. II-18),[2,3] with a HI – LO difference z-score of 2.801 (p = .003), and an effect size consistent with that found with the microelectronic random source.

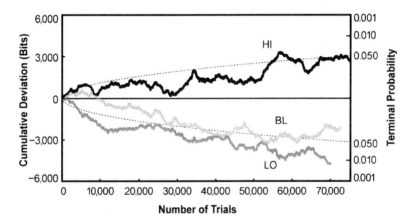

Fig. II-18. Cumulative deviation record of shift register–based PseudoREG experiment.

Upon closer investigation, however, the ramp that drove the sampling mechanism in the pseudorandom system was found to introduce a previously unrecognized physical random element into the ostensibly deterministic process. The system then was modified to eliminate the ramp, and a second smaller dataset was generated, largely by one operator, who also produced extra-chance results, albeit *opposite* to intention.

At about this time, desktop computers had become readily available, and the shift-register version was dropped in favor of one based on a pseudorandom algorithm. It is amusing, and perhaps informative, to note that in the process of developing this system, the pseudorandom sequence was initially seeded by reference to a published random number table. Two members of the PEAR staff undertook to calibrate the reliability of the algorithm by methodically entering the sequence of random numbers from the table and setting the computer to generate a sequence of one, two, or three unobserved runs of 1,000 trials each. Since a single 1,000-trial run required about 20 minutes, the number of runs selected depended upon the availability of the experimental room at the time, which was the only "random element" in the process. At the end of this exercise, the composite calibration results themselves were found to display two distinct anomalous trends that, on closer examination, were significantly correlated with the identities of the staff members who had performed the tests!

A new experiment then was developed wherein the initiating seed was computed from a combination of the current laboratory time and a microsecond timer count between the set-up and start keystrokes by the operator. Referred to as the "ATPseudo" experiment, since it was run on an IBM 286-AT computer, this more substantial study comprised a total of 482 series, or nearly 1.5 million trials.[4] The overall mean shifts in the directions of intention indicated no anomalous effects in the HI – LO differences, but there now appeared an uncharacteristic concurrence of significant HI-going trends in *both* the HI and LO results, while the baselines remained well-behaved. Remote experiments with this system displayed an even more severe case of this asymmetrical cancellation of the HI and LO effects, *e.g.* highly significant HI excursions ($p \approx 0.001$), coupled with marginally extra-chance LO deviations *also in the HI direction* ($p \approx 0.034$), that reduced the

HI – LO difference to insignificance ($p \approx 0.90$). A battery of chi-squared goodness-of-fit tests[5] confirmed that this particular form of asymmetrical anomaly was well beyond chance expectation.

The possibility that the operators' awareness that the source in the ATPseudo experiment was not physically random could have influenced their performance prompted yet a fourth study. In this, termed PS-REG, trials from a standard electronic REG source and those from the pseudorandom algorithm were interspersed. While the operators knew that half the trials were random and half pseudorandom, they were blind to which trials employed which source. Once again, the results were paradoxically ambiguous. Now *no* apparent effect was evident in the data taken from the physical source, but a marginally significant positive effect was noted in the pseudorandom HI – LO data!

Clearly, much more comprehensive results need to be accumulated to resolve this dilemma, which carries evident major theoretical implications. If it can ultimately be established that strictly deterministic sources are intrinsically impenetrable to consciousness-correlated anomalous responses, the proposed phenomenological models can be circumscribed to focus on the non-linear and chaotic dynamics of the physical noise generation processes, *per se.* If, on the other hand, genuinely pseudorandom sources can also be shown to manifest anomalous effects, even if in structurally different patterns, the epistemological search will need to dig more deeply into the nature of objective as well as subjective information, and their interactions with one another. Meanwhile, the paradox remains unresolved.

References

[1] John Von Neumann. In D. E. Knuth. *Seminumerical Algorithms*. 2nd Ed. In *The Art of Computer Programming*, Vol. 2. Reading, MA: Addison Wesley, 1981.

[2] Robert G. Jahn, Brenda J. Dunne, and Roger D. Nelson. "Engineering anomalies research." *Journal of Scientific Exploration, 1*, No. 1 (1987). pp. 21–50.

[3] Robert G. Jahn and Brenda J. Dunne. *Margins of Reality: The Role of Consciousness in the Physical World*. San Diego, New York, London: Harcourt Brace Jovanovich, 1987. Reprinted: Princeton, NJ: ICRL Press, 2009.

[4] Robert G. Jahn, Brenda J. Dunne, Roger D. Nelson, York H. Dobyns, and G. Johnston Bradish. "Correlations of random binary sequences with pre-stated operator intention: A review of a 12-year program." *Journal of Scientific Exploration, 11*, No. 3 (1997). pp. 345–367.

[5] George E. P. Box, William Gordon Hunter, and J. Stuart Hunter. *Statistics for Experimenters: An Introduction to Design, Data Analysis, and Model Building*. Wiley Series in Probability and Mathematical Statistics. New York: John Wiley & Sons, 1978.

7

MURPHY THE MACROSCOPIC

Whenever you can, count.
— Francis Galton[1]

Early in 1979, several months before the PEAR lab was formally established, the authors had an opportunity to visit the Museum of Science and Industry in Chicago. There they noticed and admired a large random mechanical cascade device, modeled after the well-known Galton Desk design.[2] Originally conceived by Francis Galton as an "Instrument to Illustrate the Principle of the Law of Error or Dispersion," it has since come to be a common demonstration of the development of random Gaussian distributions by the compounding of a multitude of binary events. As we stood before the apparatus and playfully attempted to encourage it to shift its distribution of cascading marbles to the right, we were amused and intrigued by the clearly right-shifted distribution it produced in response. During this time a group of school children looked on, listening in disbelief as their instructor, whose back was to the device, explained how it would *always* generate a properly centered normal curve. We decided on the spot that we needed to have such a machine in our new laboratory, and shortly thereafter we designed and built a version of it in our own engineering school machine shop.

Originally, this had seemed like a relatively simple task, but it

Galton Board on display at the Galton Laboratory, University College, London

actually took the better part of three years to complete, and presented an incessant sequence of technical challenges, reminiscent of Murphy's Law that "Anything that can go wrong, will." As a result, the device acquired the affectionate nickname of "Murphy." It even surpassed that law by demonstrating a few things that couldn't possibly go wrong, such as occasionally slicing some of the balls in half! There was one memorable occasion when after our technician had spent several months unsuccessfully trying to design a funnel system that would

Random Mechanical Cascade

preclude the balls jamming in the bin counters, he was given warning that if the problem wasn't solved more expeditiously "heads would roll." As this message was being transmitted, the platform supporting the ball distribution suddenly collapsed and all of the balls that were in the collecting bins crashed loudly to the floor of the storage reservoir, terrifying the poor man!

While our various microelectronic REGs, which permitted the rapid accumulation of large bodies of data, ultimately became the workhorses of the PEAR laboratory, it was Murphy, the random mechanical cascade (RMC), that turned out to be its most popular experimental device, and ultimately, its best public relations representative. Over the years, numerous TV producers were fascinated to film Murphy for their programs, and he even appeared on the front page of the *New York Times*.[3] In addition, many of the countless school children who visited PEAR used it as a model for designing their own experiments in probability. More than any of our other devices, Murphy took on a distinctly anthropomorphic character, and our operators usually addressed him by name. On one occasion when the machine was down for repairs for a few days, one of his operators actually sent him a "get well" card.

Our earliest REG results had clearly posed the categorical question of whether similar phenomena could be demonstrated using a broader range of random processors, in particular with devices of macroscopic scale, and Murphy provided an ideal opportunity to address this. Ten feet high and six feet wide, the machine was mounted on a wall in the reception area of the laboratory, facing a comfortable couch. In operation, 9000 precision-cast polystyrene balls, ¾" in diameter, trickled downward from an entrance funnel into a quincunx array of 330 nylon pegs, also of ¾" diameter, mounted on 2¼" centers. The balls bounced in complex random paths through the array, colliding elastically with the pegs and with other balls, ultimately accumulating in nineteen parallel collecting bins across the bottom. The fronts of the peg chamber and the collecting bins below it were made of transparent plastic sheets so that the cascade of balls and their developing distributions of bin populations were visible as feedback to the operators. After considerable empirical modifications to determine appropriate combinations of peg spacing, ball inlet arrangement, and material properties, the resulting distribution of ball populations in the collecting bins could be tuned to a good approximation of a Gaussian distribution (*cf.* Fig. II-19).

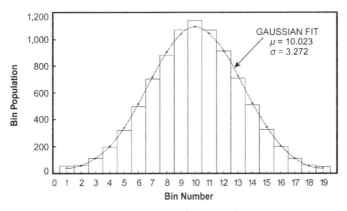

Fig. II-19. RMC bin populations.

The entrance to each collecting bin was equipped with a photoelectric sensor that detected and recorded the arrival of each ball, and the growing populations of all bins were displayed on LED counters below each bin, and graphically on a computer terminal screen. The disposition of each of the 9,000 balls in every run was recorded on-line in an appropriately coded computer file that later could be accessed to yield a faithful reproduction of the complete history of all of the bin fillings for more detailed study, or to calculate statistical properties of the terminal distributions. In addition, a photograph of the distribution and LED-displayed bin counts was taken after every run. As with all PEAR experimental devices, extensive calibrations were performed to provide background statistical data and to explore possible sensitivities to temperature and humidity, which were routinely measured and recorded before each run.

The tripolar experimental RMC protocol called for the operator, seated on the couch approximately eight feet from the machine, to attempt to distort the distribution of balls in the bins toward the right or higher numbered bins (RT), or to the left or lower numbered bins (LT), or to generate baselines (BL) with no conscious intention. These efforts were interspersed in concomitant

Random Mechanical Cascade with operator and friends

sets of three runs, each lasting approximately twelve minutes. The hundreds of experimental data sets thus obtained displayed similar anomalies in their overall concatenations to those achieved in the REG studies, including strongly operator-specific patterns of achievement. Detailed tabulations and cumulative deviation and structural graphs of the results can be found in a number of the archival references.[4–8] Here we include only the cumulative deviation plots of all data acquired in an extended sequence of these experiments (Figure II-20).

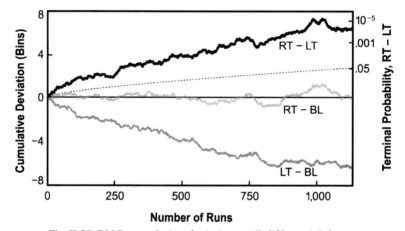

Fig. II-20. RMC cumulative deviations: All differential data.

Unlike the REG experiments, wherein theoretical baselines confirmed by calibration were available for comparison with the operator-generated data, the internal mechanics of the RMC were too complex to submit to detailed theoretical prediction. This forced us to utilize a differential criterion based upon comparison of the empirical means of the RT and LT distributions with the local baseline of the same experimental set. This strategy had the advantage of minimizing any spurious effects of short- or long-term drift in the machine operation, but introduced the confounding possibility that an operator might inadvertently influence the empirical baseline distribution as well.

And indeed, as we examined the overall cumulative deviation graphs, plotted as RT − LT, RT − BL, and LT − BL differences, an intriguing secondary anomaly appeared. Whereas it was abundantly evident that the overall RT − LT mean separation was statistically highly significant (ε_μ = 1.93, z_μ = 3.89; p_μ = 5 × 10^{-5}), it also displayed a curious asymmetry in the LT direction. Namely, virtually all of the compounding RT − LT anomalous deviation was attributable to the LT − BL separation alone; the RT and BL evolutions were statistically indistinguishable!

For some time we attempted to resolve this asymmetry empirically: operators changed their positions on the couch, closed and opened laboratory doors, and one mounted a mirror on the facing wall and observed the reflected runs. One even stood on his head, but to no avail! It was several years later, in the course of our study of gender differences, described in Chapter II-13, that it was discovered that this propensity was entirely attributable to the tendency of many of the female operators to produce baselines that were strongly shifted in the right-going direction, thus producing results that showed a significant deviation in their LT − BL efforts, but a null result in the RT − BL. A similar gender-related trend subsequently was found to prevail in several other PEAR experiments.

As in the REG experiments, the total number of runs conforming to the intended direction to any degree was found to be considerably higher than the chance prediction, so that once again we concluded that the overall patterns of anomalous shifts of the mean were being compounded from an accumulation of small individual effects. We also again observed some operator-specific dependencies of the results on the secondary parameters of the experiment, such as the time of day, the volitional vs. instructed assignment of run order, or whether the LED count display was on or off.

Perhaps of higher importance, however, was the similarity of many of the individual operator cumulative deviation patterns with those they demonstrated in the microelectronic REG experiments. Despite their inherently stochastic character, the evident gross similarities of their "signatures" had major implications for experimental design and theoretical modeling. Namely, although the observed anomalous effects were clearly operator-specific and in many cases condition-specific, they appeared not to be nearly so device-specific, a feature later confirmed over a much wider range of physical processes, scales, and energies. Thus, once again, it appeared that any direct influence of operator consciousness on these widely different physical processes, *e.g.* the flow of electrons in the REG noise diode, or the cascade of balls in the macroscopic RMC, was less likely to be a direct dynamical mechanism than a more holistic interaction with the statistical information common to both these systems.

Finally, we might note that although the RMC differed substantially from the REG devices in its scale and physical process, it retained a certain quasi-digital character in the manner in which it generated information. Specifically, each falling ball, upon collision with a peg, might be diverted either to the right or to the left, and it was the compounding of these binary right/left options that primarily determined the terminal distributions in the bins. To be sure, in this machine the binary right/left probabilities were not simply .50/.50, since the balls did collide with one another as well, and therefore their subsequent trajectories were not at all uniform, but nonetheless, a synthetic binary quantification could be, and actually was, imposed in the analyses.

A further step in tracking the ubiquitousness of operator-related anomalies, therefore, was to apply similar protocols to physical systems that were yet more analogue in character, even to those whose central random processes and outputs lent themselves to continuum representation. All of these experiments utilized similar tripolar protocols to those followed for their digital counterparts, and from this array of studies, a brief review of which is included in Chapter II-11, we broadened our conclusion that the specific character of the physical random sources employed was *not* a primary correlate of their anomalous responses.

When the PEAR laboratory closed in 2007, perhaps the most emotionally poignant moment was Murphy's disassembly. He had played a vital role in our program, both in the valuable data he had produced and in his contribution to the laboratory's physical and subjective ambience. His Princeton career has been insightfully celebrated in a charming article by Graham Burnett in *Cabinet: A Quarterly Journal of Art and Culture*[9] that delves rather deeply into the metaphysical implications of these experiments, and in a brief review in *EdgeScience* magazine.[10]

References

[1] Sir Francis Galton. Quoted in James R. Newman, "Commentary on Sir Francis Galton," in *The World of Mathematics*. New York: Simon & Schuster, 1956. p. 1169.

[2] Sir Francis Galton. *Natural Inheritance*. New York and London: Macmillan, 1894. pp. 63–65.

[3] Benedict Carey. "A Princeton Lab on ESP Plans to Close Its Doors." *New York Times*. February 6, 2007.

[4] Brenda J. Dunne, Roger D. Nelson, and Robert G. Jahn. "Operator-related anomalies in a random mechanical cascade." *Journal of Scientific Exploration*, *2*, No. 2 (1988). pp. 155–179.

[5] Brenda J. Dunne. "Gender differences in human/machine anomalies." *Journal of Scientific Exploration*, *12*, No. 1 (1998). pp. 3–55.

[6] Brenda J. Dunne and Robert G. Jahn. "Experiments in remote human/machine interaction." *Journal of Scientific Exploration*, *6*, No. 4 (1992). pp. 311–332.

[7] Brenda J. Dunne, York H. Dobyns, Robert G. Jahn, and Roger D. Nelson. "Series position effects in random event generator experiments." *Journal of Scientific Exploration*, *8*, No. 2 (1994). pp. 197–215.

[8] Robert G. Jahn and Brenda J. Dunne. "The PEAR proposition." *Journal of Scientific Exploration*, *19*, No.2 (2005). pp. 195–246.

[9] D. Graham Burnett. "Games of Chance." *Cabinet: A Quarterly Journal of Art and Culture*, *34*, Summer 2009. pp. 59–65.

[10] Robert G. Jahn and Brenda J. Dunne. "The effects of human intention on a machine named Murphy." *EdgeScience*, No. 4, July/September 2010. pp. 14–16.

8

RIDE 'EM ROBOT

There are some things so serious,
you have to laugh at them.
— Niels Bohr[1]

The anthropomorphic appeal that made "Murphy" so popular with our operators was invoked in another PEAR experiment that featured a small robotic vehicle driven to meander over a circular tabletop by an onboard microelectronic REG specifically adapted to that purpose. To enhance its attractiveness, the electrical and mechanical components of the robot were encased in a 15-cm dome-shaped housing, somewhat resembling a miniature Zamboni machine, with an endearing toy frog perched in a driving position. Originally commissioned to extend the search for more engaging dynamic feedbacks that might enhance operator resonance with the experimental tasks, this whimsical device fascinated generations of operators, schoolchildren, other visitors, laboratory interns, and staff for decades, while providing instructive data for several archival publications and presentations at professional conferences.[2–4]

The concept for this experiment was originally proposed by a French scholar, René Péoc'h, who himself had appropriated robotic inventions, termed "tychoscopes," created by two of his colleagues, P. Janin[5] and R. Tanguy,[6] for experiments involving young animals.[7] In one

REG-driven PEAR Robot

131

version, newborn chickens were imprinted on a remotely driven robot by placing them in its presence for one hour every day for six days after their birth. They were then separated from this "mother" in a transparent cage from which they could see the moving robot, but were unable to follow it around. The device was found to spend significantly more time on the half portion of its propagation surface closest to the chicks, compared to its motion when the cage was empty. In the presence of a control group of chicks that had not been conditioned to adopt it as their mother, the robot moved in its normal random motion. In subsequent experiments with light-deprived chicks and baby rabbits, the animals succeeded in attracting the robot to them when a candle was placed on top of it. From such studies, Péoc'h established the capacity of these animals to affect the trajectory of the robot to their biological advantages, by some anomalous means.[8]

Our extension of these techniques to involve human operators addressed the hypothesis that an anthropomorphic resonance with the robot behavior would enhance anomalous alterations of its random trajectory, with corresponding departures of the digital output of the REG unit directing it. For the formal experiments, the device was deployed on a 48"-diameter circular table in one of our principal experiment rooms, with the operators seated adjacent to the table, but having no contact with it. The first protocol consisted of alternating efforts on the part of the operators to direct the robot to two opposite positions on the circular table, labeled 0° and 180°; each such effort was termed a "run," and two successive alternative efforts, a "set."

The on-board mechanism driving the device comprised two independent, battery-powered clock motors, each controlling one wheel of the robot. These in turn were instructed by the REG unit to drive the wheels by a sequence of various incremental amounts, thereby accomplishing a random array of forward translations and clockwise or counter-clockwise rotations of the vehicle. From a prescribed initial position and direction at the center of the table, the device executed a two-dimensional stochastic

Robot operation

trajectory, eventually reaching the table edge. Once switched on, the robot motion began with a rotation, after which it randomly alternated forward translations with subsequent rotations. When the robot was due to start a translation, the drive system compared a particular binary composition with two values that separated the theoretical distribution into three equal segments, on the basis of which it went forward for four, five, or six steps, with equal probability.[2] After completing that translation, the robot then rotated clockwise or counter-clockwise, based upon whether the next binary composition was positive or negative, or zero, in which case it proceeded through another translation whose distance was computed from the following bit sequence. In its rotation mode, the robot examined a succession of binary digits to determine whether to continue its rotation or to stop, with a probability of approximately one in ten to stop after any given bit sequence.

An LED on top of the robot dome allowed an overhead digital camera to track and record the motion of the device. The x and

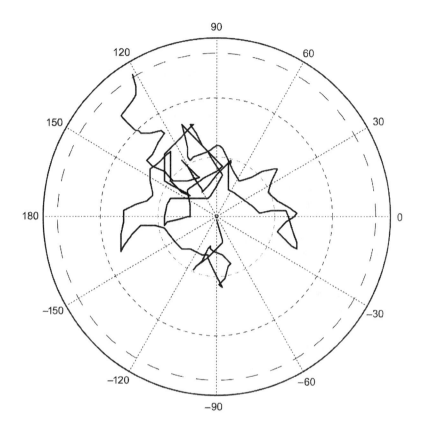

Fig. II-21. Typical Robot trajectory.

y coordinates of this LED position were recorded three times per second, along with the time of measurement, and these files were the basis for all subsequent analyses of the robot trajectories. Figure II-21 displays an example of a digitized robot trajectory extracted from such a data file.

Statistical assessment of the overall results of these experiments, and of their structural characteristics, were inescapably complicated by the irreducible vagaries in the robot's mechanical performance, and by the circular geometry of its dynamics. Detailed calculations and tabulations have been presented in the parent publications[2,3] from which one summary table has been

extracted simply to illustrate the extent of the structural anomalies appearing in the data (Table II-5). Specifically to be noted
are that although the overall database of this experiment, which
compounded to 1120 sets contributed by 87 operators (41 females
and 30 males), was statistically unimpressive, there was a clear
gender disparity in the results, with the females achieving a modestly significant separation in their directions of intentions, while
a significant fraction of the males tended to produce results opposite to their intentions. A smaller subset of 65 sets, produced
by 6 groups comprising several operators performing together, attained an even more substantial effect size, comparable to those
observed in several of the other REG-based experiments.

TABLE II-5

Summary of All Robot Exit Angle Experiments, 0° vs. 180°

Subset	# Ops.	# Sets	z (p_z)	ε	χ^2_{op} (p_x)	# Ops. +
All	87	1120	.957 (.169)	.0286	85.561(.524)	48
Females	41	567	1.878 (.030)*	.0789	50.743(.142)	27*
Males	30	416	−1.164 (.880)	−.0571	23.011(.815)	12[†]
Co-Operators	10	72	−.298 (.617)	−.0351	5.444(.860)	4
Groups	6	65	1.738 (.041)*	.2123	6.363(.384)	5*
Calibrations	—	348	.810 (.209)	.0434	—	—

KEY

Ops: Numbers of individual operators, operator pairs, or groups contributing to the databases;

Sets: Numbers of 0°, 180° paired sets performed;

z: Statistical z-scores;

p_z: One-tailed probabilities of z-scores against chance expectations ($*$ denotes significance at $p_z \leq .050$);

ε: Equivalent effect sizes, computed as $z / \sqrt{\text{# Sets}}$;

χ^2_{op}: Statistical chi-squared calculations over individual operator z-scores; i.e., $\sum_{ops} z^2$ to be compared with the number of degrees of freedom (number of operators) to estimate the probabilities of the z distribution structures compared to chance expectations, p_χ;

Ops. +: Number of operators exceeding chance mean expectations in the intended directions ([†] significantly less than chance expectation).

In subsequent rounds of experiments, operator efforts were concentrated on a "time-of-flight" mode, wherein were compared the residence times of the robot on the table for alternating efforts to hasten its exit, or to prolong its excursion.[4] Time differences were compiled from successively interspersed long-intention and short-intention trials performed in a single period of operation, to minimize any effects of secular drift in the mechanical performance of the robot itself. Owing to the small size of this database, the statistical results were only modest, but the anomalous effect sizes were comparable with those of our best other REG-based experiments.[3]

PEAR Robot and friends

Briefly summarized, these studies verified the ability of individuals or groups of operators preferentially to direct the robot to progress from its initial placement at the table center to pre-specified segments of the table edge; to spend greater, or lesser, time in transit from the center to the edge; or to cover greater or less total trajectory distance in such transits. Gender disparities and serial position effects were evident and, as in most of our other experiments, these and the other modest results compounded statistically in small increments over large databases to a degree sufficient to support several consequential inferences:

1) the intrinsic feedback to the operators provided by the chaotic motion of this whimsically attractive device established sufficient operator/machine resonance to elicit various significant anomalous responses, comparable to those found in the bench-mark REG studies;

2) the consistency of this structural pattern of results with those obtained using substantially different equipment and pro-tocols reinforced the generic character of such phenomena that eventually may contribute to a useful comprehensive model for their representation and interpretation; and

3) possibilities for beneficial human/machine systems em-bodying consciousness-resonant robotic components should not be dismissed out of hand.

References

[1] Niels Bohr. As quoted in Abraham Pais, *The Genius of Science: A Portrait Gallery.* Oxford, New York: Oxford University Press, 2000. p. 24.

[2] Robert G. Jahn, Brenda J. Dunne, David J. Acunzo, and Elissa S. Hoeger. "Response of an REG-driven robot to operator intention." *Journal of Scientific Exploration, 21,* No. 1 (2007). pp. 27–46.

[3] Robert G. Jahn, Brenda J. Dunne, David J. Acunzo, and Elissa S. Hoeger. "Response of an REG-Driven Robot to Operator Intention." Technical Report PEAR 2006.03. Princeton Engineering Anomalies Research, Princeton University, School of Engineering/Applied Science, Princeton, NJ. November 2006 (43 pages).

[4] Robert G. Jahn, Ezdean B. Fassassi, John C. Valentino, and Elissa S. Hoeger. "Random Robot Redux: Replications and Reflections." Technical Note PEAR 2008.01. Princeton Engineering Anomalies Research, Princeton University, School of Engineering/Applied Science; and ICRL #07.7, International Consciousness Research Laboratories, 211 N. Harrison St., Suite C, Princeton, NJ 08540-3530. March 2008 (43 pages).

[5] Pierre Janin. "The tychoscope: A possible new tool for parapsychological experimentation." *Journal of the Society for Psychical Research, 53,* No. 804 (July 1986). pp. 341–347.

[6] Roger Tanguy. "Un Réseau de Mobiles Autonomes pour l'Apprentissage de la Communication." Doctoral thesis, Université Paris 6, 2 décembre 1987.

[7] René Péoc'h. "Chicken imprinting and the tychoscope: An Anpsi experiment." *Journal of the Society for Psychical Research, 55* (1988). pp. 1–9.

[8] René Péoc'h. "Psychokinetic action of young chicks on the path of an illuminated source." *Journal of Scientific Exploration, 9,* No. 2 (1995). pp. 223–229.

9

PENDULUM PLAY

Here's a new day. O Pendulum move slowly!
— Harold Monro[1]

The similarity of the anomalous results achieved with the large-scale mechanical RMC apparatus to those of the microelectronic REG devices begged the broader empirical question of just how ubiquitous such phenomena could be shown to be with respect to their underlying sources of randomness. As described in previous chapters, several pertinent experiments were performed addressing substantially differing physical processes, but all of these were essentially digital or discrete in nature, with positive or negative binary increments in the experimental measures as the target of the operator's intentions. To increase the generality, these explorations were extended into the analog domain via experiments that had potentially greater sensitivity to operator interaction because of the continuously variable nature of their measurables, and/or by the incorporation of aesthetic qualities that might enhance the operator/machine resonance.

As a first attempt, we undertook to design and construct a classical linear pendulum suitably instrumented to provide precise measurement of its dynamic performance as well as to give stimulating feedback to operators. Of the many possible implementations, a free-swinging but enclosed rod-and-bob pendulum was chosen for development, with the damping rate selected as the primary measurable.[2] What first appeared to be a relatively straightforward project turned out to be a formidable technical task that challenged the ingenuity of several members of our staff, and ultimately took several years to bring to credible operation. The extraordinary sensitivity of this device, the extensive calibrations and adjustments required to insure a reliable pattern of performance, the complexities of an attractive feedback system that could track and display the subtle analog fluctuations, and the determination of appropriate analytical strategies for converting the differential outputs to a form suitable for statistical assessment, all conspired to make this one of the most challenging experiments in the PEAR repertoire. Notwithstanding, after an arduous period of preparation, it finally achieved a sufficiently reliable status to permit formal data collection.

Briefly, the ultimate system consisted of a two-inch diameter crystal sphere suspended to swing on a thirty-inch-long fused silica rod, all enclosed in a clear acrylic box. An interior computer-controlled mechanical release system ensured identical starting conditions for each run of one hundred swings of the pendulum bob, undisturbed by any human handling. Photoelectric detectors recorded with microsecond accuracy the intervals of passage of a double knife edge mounted on the rod near the bob past a high-speed counter, from which nadir velocities of each swing could be computed with high precision. Each hundred-swing run lasted a total of about three minutes, during which operators attempted to minimize the rate of swing damping, *i.e.* to sustain the swing amplitude (HI); to increase the damping rate, *i.e.* to reduce the swing amplitude (LO); or to take an undisturbed baseline (BL). In addition, sensitive instrumentation was incorporated to record the temperature, humidity, and barometric pressure automatically over the course of each run, all of which were included in the index file. A proximate sequence of BL, HI, LO, or BL, LO, HI, runs comprised a *set*, collections of which constituted the databases for each operator's efforts in given parametric categories of the experiment.

It was presumed that the random dynamical contribution to the pendulum motion derived primarily from the turbulent airflow around and in the wake of the bob and support rod (somewhat akin to that engendered by a baseball knuckleball pitch), although friction in the rod-bearing suspensions also may have contributed

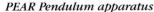

PEAR Pendulum apparatus   **Operator with Pendulum**

some stochastic irregularity to the bob motion, as well. In any case, this machine clearly qualified as macroscopic in scale and ostensibly analog in character, even though the velocity-measuring system eventually digitized its output for recording purposes.

Since theoretical modeling of idealized pendulum function could provide only a rough approximation to the precise empirical measures taken of the complex real system, experimental data could be assessed only against a background of calibrations that characterized the performance in the absence of operator interactions. These were generated automatically at regular intervals in sets of 27 runs, some in sessions typically beginning at 2:00 AM, and others during the day, to determine whether the activity of people in the laboratory could detectably influence pendulum performance. The results of these two categories of calibrations were found to be statistically indistinguishable and were subsequently combined. To assess their distribution characteristics and confirm the validity of the statistical processing, these calibration data were arbitrarily assigned to the three pseudo-intention categories, then processed as if they were real experimental data taken in 9-set "series," and comparisons were made between the artificially defined "HI" and "LO" results, and both of these then compared with the "BL." This random assignment procedure was used to construct 600 artificial series t-scores. A goodness-of-fit comparison of these with the appropriate theoretical Student t-score distributions yielded χ^2 = 11.964, based on 13 degrees of freedom, with a corresponding chance probability of 0.531. Thus, although some session-to-session changes due to atmospheric effects were detectable in the calibration mean and standard deviation values, these were normalized correctly by the within-set differential analysis, thereby qualifying this distribution as an appropriate basis of comparison with the active experimental data.

One option for operator feedback took the form of axial illumination of the bob by a color-coded lamp reflective of the local damping rate, causing it to glow red or amber for less damping (HI), or green to blue for more damping (LO). This particular feedback option, used in conjunction with a numerical display of the digital data, turned out to be the most popular for the operators.

As in most of our other experiments, the primary variable of interest was the difference between the HI and LO intentions, although a number of secondary parameters were also explored, including a subset of trials conducted with the operators physically situated in remote locations. In this remote protocol, the pendulum was programmed to generate data automatically in session-length blocks beginning at specified times, with runs spaced at five-minute intervals. Operators reported the order of the HI and LO intentions after the data had been generated and recorded, but before receiving feedback on the results. Also addressed was the instructed vs. volitional determination of the run order. Although all sets began with the BL condition, comparison with which determined the feedback provided to the operator during the subsequent intentional efforts, at the beginning of each 27-run session the order of HI and LO efforts could either be assigned by a pseudorandom algorithm or selected by the operator. The operator was also permitted to choose the desired form of feedback, *e.g.* the changing color of the illuminated pendulum bob accompanied by a digital display on the computer screen, or the digital display alone. In the remote experiments, of course, no run-by-run feedback was available. The composite results, along with those of the various secondary subsets, are summarized in Table II-6.

Over the 915 local data sets generated by 42 operators (16 male and 20 female), the overall HI – LO differences were statistically unremarkable. There were significant differences, however, between both the volitional/instructed and feedback options, as well as between the male and female performances, most of which tended to cancel each other in the composite results. Figure II-22(a,b) displays the cumulative deviation traces for the instructed and volitional data, where the contrast in performance was particularly striking.

Both males and females produced positive HI – LO results in the instructed protocol and negative results in the volitional, although in both conditions the males outperformed the females, significantly so in the instructed protocol. The male efforts were statistically significant in both the color and digital feedback conditions. These results were confirmed by specific subset data analyses and more comprehensive chi-squared and ANOVA data treatments.[2] The gender differences were consistent with those observed in other PEAR experiments, where the male operators succeeded in achieving positive correlations with their stated intentions and the females generated results that were asymmetrical relative to intention and displayed larger variances than those of the males,[3] an asymmetry reflected here in the differences between the combined HI and LO efforts and the BLs.

As in our other experiments, the pronounced gender differences that were observed in the local efforts were not found in the remote data. The composite results of the 630 remote sets

TABLE II-6

Pendulum Damping Rate Database: HI – LO and INT – BL
Comparisons by Major Parameters

Subset	# Sets	HI – LO	SD	HI – LO z-score	Prob.	INT – BL z-score	Prob.
All Local Data	915	1.294	100.8	0.388	0.349	1.245	0.107
Volitional	421	–7.470	101.9	–1.502	(0.067)	1.207	0.114
Instructed	494	8.762	99.5	1.953	0.025*	0.572	0.284
Color/Digital Feedback	666	5.004	97.4	1.325	0.093	0.993	0.160
Digital Feedback Only	249	–8.630	109.2	–1.244	(0.107)	0.751	0.226
Female Data	609	–3.448	106.4	–0.799	(0.212)	0.664	0.253
Male Data	306	10.730	88.2	2.118	0.017*	1.353	0.088
Female Volitional	313	–9.520	109.1	–1.540	(0.062)	1.054	0.146
Male Volitional	108	–1.528	77.5	–0.205	(0.419)	0.592	0.277
Female Instructed	296	2.973	103.2	0.495	0.310	–0.128	(0.449)
Male Instructed	198	17.416	93.1	2.606	0.005*	1.238	0.108
Female Color Feedback	377	1.765	103.6	0.331	0.370	0.317	0.376
Male Color Feedback	289	9.229	88.6	1.764	0.039*	1.262	0.104
Female Digital Feedback	232	–11.918	110.5	–1.636	(0.051)	0.656	0.256
Male Digital Feedback	17	36.244	79.5	1.760*	0.039*	0.522	0.301
All Remote Data	630	2.456	92.4	0.667	0.252	0.642	0.260
Female Remote	161	9.256	104.9	1.116	0.132	0.136	0.446
Male Remote	469	0.121	87.7	0.030	0.488	0.682	0.248
All Data	1545	1.767	97.5	0.713	0.238	1.367	0.086

Key

Sets: Numbers of BL, HI, LO, or BL, LO, HI, proximate sets of runs performed under particular parametric conditions;

HI – LO: Differences in commensurate values of the measurables for HI and LO intention efforts;

SD: Standard deviations, in appropriate units;

HI – LO z-score: Statistical z-scores of HI – LO differences;

INT – BL z-score: Statistical z-scores of differences of intentional runs – baselines;

Prob.: One-tailed probabilities of z-scores against chance expectations.

* indicates statistical significance ($p < .05$).

Numbers in parentheses ( ) denote effects opposite to intention.

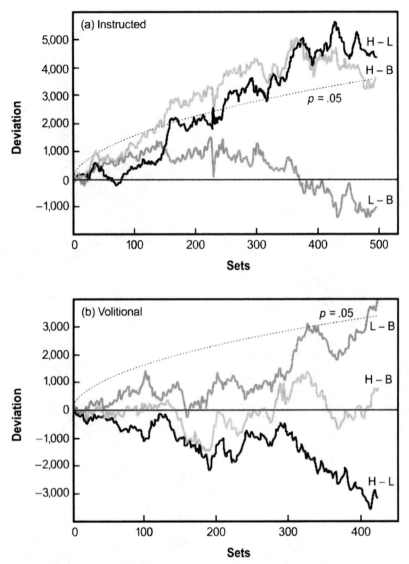

Fig. II-22. Cumulative deviation graphs of Pendulum data: a) Instructed trials; b) Volitional trials.

had an overall null result, but this turned out to be confounded by the data of one particular operator, which comprised nearly half the total database and displayed a consistent negative trend that severely depressed the overall remote results (Figure II-23a).

In contrast, the combined results of the 11 other operators were positive and marginally significant ($p = 0.048$), with an effect size considerably larger than that of the local database (Figure II-23b).

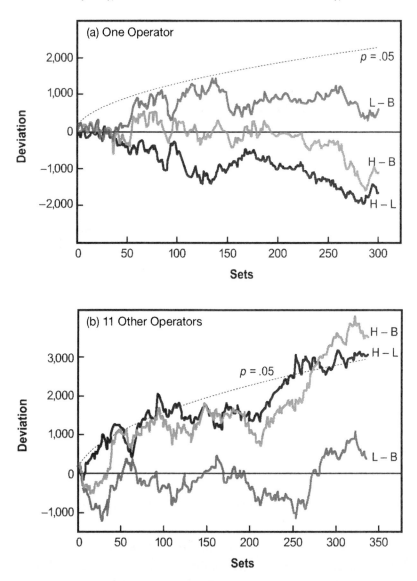

Fig. II-23(a,b). Cumulative deviation graphs of remote Pendulum data: a) One particular operator; b) Other 11 operators.

Thus, despite our ingoing aspirations that the analogue character and the attractive, possibly even hypnotic, nature of the feedback would enhance the anomalous effects, the results of this pendulum experiment proved only comparable in scale to those obtained on the Benchmark digital devices and on the large-scale RMC. Once again, despite their idiosyncratic character, the lack of sensitivity of the phenomena to the particular kind of random source being addressed was reinforced, and taken in conjunction with many other similar empirical indicators, emphasized the need for any viable theoretical representations indiscriminately to span a broad range of random devices and processes.

References

[1] Harold Monro. "Living." In Alida Monro, ed. *Collected Poems* (London: Duckworth, [1933] 1970). p. 13. (published on <guardian.co.uk> at 12.49 GMT on Monday 7 December 2009.)

[2] Roger D. Nelson, G. Johnston Bradish, Robert G. Jahn, and Brenda J. Dunne. "A linear pendulum experiment: Effects of operator intention on damping rate." *Journal of Scientific Exploration, 8,* No. 4 (1994). pp. 471–489.

[3] Brenda J. Dunne. "Gender differences in human/machine anomalies." *Journal of Scientific Exploration, 12,* No. 1 (1998). pp. 3–55.

10

DOUBLE-SLIT DISPARITY

We choose to examine a phenomenon
which is impossible ...
— Richard Feynman[1]

In 1996, Stanley Jeffers, a physicist colleague based at York University, Toronto, proposed an application of the classical Young's double-slit optical diffraction experiment illustrated below,[2] in which participants would be asked to reduce the interference fringe contrast, as a possible discriminator of potential modes of anomalous human/machine interactions.

The experiment was originally conceived as a means of testing whether statistical anomalies such as those observed in the PEAR studies derived from an ability of the human operator to observe and collapse a quantum-mechanical wave function by extra-sensory means, or from an ability to select from alternative

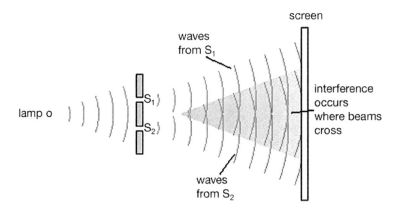

Fig. II-24. Young's Double-Slit experiment.

outcomes. Specifically, human operators were invited to "visual-ize" monochromatic light passing through a double slit, *prior* to its registration as an interference fringe pattern by an optical detec-tor. It was postulated that such extra-sensory observation would manifest as a measurable departure of the interference pattern from theory or from calibration due to premature wave-function collapse.

The double-slit interference pattern was generated by a low power He-Ne laser illuminating a commercially available stainless steel disc in which two slits were cut, 10 mm wide and 10 mm apart. A stepper motor rotated the disc to chop the beam close to the laser aperture, effectively blocking the beam for two out of every three seconds. The resulting interference fringes were detected by a linear diode array, with the response during the in-tervals with the beam blocked serving as the background to be subtracted from the response during the intervals with the beam un-blocked. The diode array responses during the blocked and un-blocked phases were digitized and transmitted to a computer for analysis. (See Figure II-25.)[3]

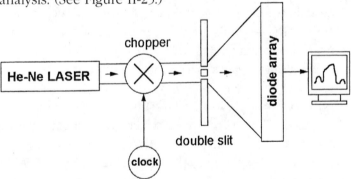

Fig. II-25. Schematic of Jeffers' Double-Slit apparatus.

An operator session consisted of a series of 41 runs, alternately designated "active" or "inactive," comprising 11 and 10 trials each, respectively, wherein a trial consisted of a one-second exposure. During the active runs, the interference pattern and an analogue indicator of the contrast (a vertical bar of variable height) were displayed to the operator and updated immediately following the close of each trial. During the inactive runs, the display was blanked and the operator remained present but did not try to influence the results.

Seventy-five such series carried out at York University showed no statistically significant effect of operator intention ($z = -0.481$). After these results were duly reported,[4] Prof. Jeffers graciously offered to transfer his equipment to the PEAR laboratory, where it was subsequently refined and repaired, and a pre-specified dataset of 20 series were performed, following the same protocol. In contrast to the null York results, the small PEAR database compounded to a terminal z-score of precisely 1.654 ($p = 0.050$), with an effect size comparable to that achieved on the much larger database benchmark REG studies.[5] Both of these results are displayed in Figure II-26 in the form of cumulative deviation traces.

Technical and theoretical details of these twin experiments are well covered in the two cited references and will not be reprised here. What may be most important to note, however, is the indication these results provide of the pertinence of subjective factors in the achievement and replication of such anomalous effects, with the obvious implications thereof for experimental design and performance, as well as for proposition of theoretical models. Prior to participation, in the experiments conducted at York University, the operators were presented with a technical description of the apparatus and the quantum mechanical process that it entailed. They were then instructed to imagine that during the active runs they could identify (by extra-sensory means) the path of the light beam near to the double slit and told that success at this task would be reflected in a less-well-defined interference pattern and a corresponding reduction in the contrast reported by the analogue detector. In contrast, the PEAR operators were simply invited to participate in a "new experiment," where their primary task would be to try to make the analogue indicator bar remain as low as possible. It is our suggestion that the contrast between the technically detailed and dispassionate approach of

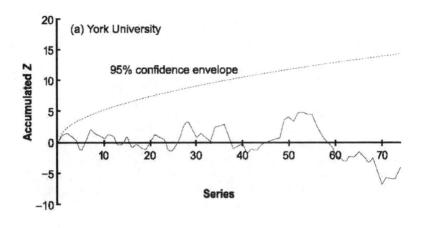

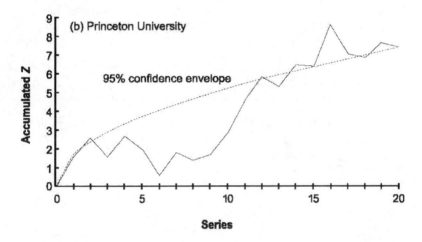

Fig. II-26. Accumulated *z*-score ($z\sqrt{N_{Series}}$) from Double-Slit experiment:
a) York University; b) PEAR laboratory.

the York investigator in presenting the task to his operators, and our own more affable and participatory tone in their oversight, may be consequential, *i.e.* that the expectations of the operators may be a salient factor in their performances.

151

References

[1] Richard Phillips Feynman, Robert B. Leighton, and Matthew Sands. *The Feynman Lectures on Physics.* Vol. 3. Addison-Wesley: Reading, 1965. p. 2.

[2] Stanley Jeffers, R.D. Prosser, G. Hunter, and J. Sloan. Classical Electromagnetic Theory of Diffraction and Interference: edge, single- and double-slit solutions. In: *Proceedings, Waves and Particles in Light and Matter: Workshop on the occasion of Louis de Broglie's 100th Anniversary, Trani, Italy, Sep 24–30, 1992.* Alwyn van der Merwe and Augusto Garuccio, Eds. New York, London: Plenum, 1994.

[3] Michael Ibison and Stanley Jeffers. "A Double-Slit Diffraction Experiment to Investigate Claims of Consciousness-Related Anomalies." Technical Note PEAR 97006. Princeton Engineering Anomalies Research, Princeton University, School of Engineering/Applied Science, Princeton, NJ. July 1997 (11 pages). p. 3.

[4] Stanley Jeffers. Intentionality and Complementarity: What might the double-slit experiment tell us about consciousness? Presented at *Toward a Science of Consciousness II*, Tucson, Arizona, April 1996.

[5] Michael Ibison and Stanley Jeffers. "A double-slit diffraction experiment to investigate claims of consciousness-related anomalies." *Journal of Scientific Exploration, 12*, No. 4 (1998). pp. 543–550.

11

SUNDRY SORTIES

*"The time has come," the Walrus said,
"to talk of many things: of shoes — and ships —
and sealing wax — of cabbages and kings —
and why the sea is boiling hot — and
whether pigs have wings."*
— Lewis Carroll[1]

Beyond the major investments in the acquisition of very large databases in the REG and other genres of experiment described in the previous chapters, over its 28-year history the PEAR program pursued a variety of less productive projects. Some of these were themselves quite extensive investigations that appeared to produce anomalous results, but for one reason or another proved intractable in their data reduction and/or analytical demands. Others were shorter-term pilot studies searching for new candidates for more intensive and detailed research expeditions, and some addressed specific issues of strategy, technology, or protocol. Our purpose in summarizing these here is simply to provide the reader with an account of the full range and depth of such explorations in the hope of precluding unproductive duplicate research efforts, while yet stimulating new imaginative ideas.

A. Fountain Experiments

Another substantial attempt to derive an aesthetically attractive feedback from a categorically different physical source featured a glittering jet of water that projected a well-formed laminar column vertically upward from a small nozzle, eventually to collapse back upon itself in an overlay of turbulent blobs, much like the larger-scale fountains commonly seen in parks and outdoor landscaping. The height of this hydrodynamic transition from the columnar laminar jet to the chaotic turbulent blobs was, both empirically and theoretically, a statistical matter and hence tended to bounce up and down over a bounded random range dependent on the flow parameters and the nozzle dimensions. Photoelectric instrumentation tracked the oscillations of this transition interface, while operators attempted to influence it to raise or lower in accordance with their pre-stated intentions, using our standard tripolar protocol.

The plumbing system required to support this experiment was necessarily complex, involving an assortment of pumps, filters, valves, and vibration isolation couplings to preclude mechanical, acoustical, or fluid pressure variabilities from artifactually disturbing the intrinsic random patterns displayed by the distilled water jet. Likewise, the photoelectric optical system required precise adjustment and maintenance to provide reliable data while maintaining an attractive luminous display for operator feedback.

PEAR Fountain apparatus   *Operator with Fountain*

Numerous individual operators provided many tripolar data-sets on this device, achieving overall results that appeared to bear some similarity to those of a few of our other experiments, *e.g.* characteristic individual operator features, gender disparities, and other sub-structural anomalies, including asymmetric cancelling of HI – LO effects. Unfortunately, analytical complexities preclud-ed our drawing any definitive quantitative conclusions.

A variant of the basic protocol was also attempted wherein the optical detection system monitored not the transition height *per se*, but the total light scattered from the turbulent portion of the jet, as a crude index of the degree of its chaotic, luminous activ-ity. This "sparkle" mode was especially appealing to many of the operators, but also yielded no extraordinary correlations beyond the now familiar idiosyncratic structural effects. Once again, we were left with equivocal indications that this particular class of random physical source was also modestly susceptible to operator influence, but that the attractiveness of its feedback seemed not to have a major influence on overall performance.

Two other variants of hydrodynamic sources were explored during this phase of the experimental program. One was struc-tured around a jet of water projecting downward, rather than up-ward, but displaying a similar spontaneous transition from lami-nar to turbulent flow. This "down fountain," which can readily be simulated using any kitchen sink faucet, proved so sensitive to spurious disturbances, such as building vibrations, handclapping, or even human voices, that it could not be adequately stabilized to reliable baseline or calibration operation, hence was never em-ployed as a target for operator intentional experiments.

A yet more radical scheme attempted to employ a borrowed student water tunnel that could create a conventional "hydraulic jump" in its channel flow. (This phenomenon can also be readily demonstrated in the impingement of faucet outflow on a smooth sink surface.) The idea was to attempt to move the jump loca-tion upstream or downstream by operator attention alone, but the inherent instabilities in the system precluded any reliable deployments.

B. Drumbeat

At one point in our broadly cast search for better understanding of the role of feedback in stimulating anomalous human/machine outputs in REG equipment, we conceived, constructed, and operated a body of experiments wherein audio rather than visual feedback was provided. To implement this modality, the output from one of our small Johnson noise sources[2] was electronically processed to drive a drumstick mechanism that beat a large Native American drum in one of two auditory modes:

1) a regularly spaced sequence of randomly alternating loud and soft beats ("amplitude" mode); or
2) a sequence of uniform-amplitude beats of two randomly alternating spacing intervals ("rhythm" mode).

The hypothesis to be addressed in this experiment was that the operator's unconscious psyche would endeavor to insert some ordered pattern onto the auditory bit strings, perhaps akin to those first conceived by primordial human consciousness, or even by non-human species given to auditory communication. Our analysis packages searched for demonstrable disparities between

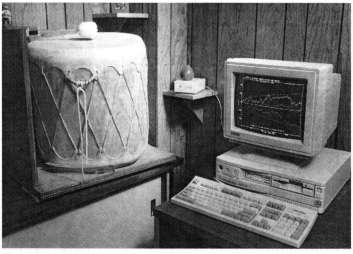

Drumbeat experiment

alternating "attended" and "unattended" output bit data. For example, we tabulated the frequencies of occurrence of particular binary "words," *e.g.* singlets (0 or 1), doublets (00, 11, 01, 10), triplets (000, 111, 001, 010, …), quartets (0000, 1111, 0001, …), and quintets (00000, 11111, 00001, …), or more elaborate mixes thereof that comprised ostensible ordered patterns. Alternatively, we performed more elaborate calculations of the total digital Shannon entropies of the binary strings (defined as: $S = \sum_i p_i \ln p_i$ where p_i denoted the frequency of occurrence of the digital elements in the particular binary figures).[3] The attended and unattended values were compared with each other, or with the corresponding theoretical expectations, using standard statistical measures such as two-sample *t*-tests, in a search for significant disparities.[4–6] The empirical results of such "word" frequency-of-occurrence comparisons are summarized in Table II-7:

TABLE II-7

_t_-Scores of (Attended – Unattended) Entropy Differences
of Words of Length 1–5, and Their Associated 1-Tailed Probabilities

Word length		1	2	3	4	5
Amplitude mode	t	1.536	1.573	1.544	1.572	1.603
	p_t	0.063	0.058	0.062	0.058	0.055
Frequency mode	t	−0.420	0.471	0.510	0.614	0.913
	p_t	0.663	0.319	0.305	0.270	0.180
Combined data	t	1.447	1.630	1.614	1.671	1.790
	p_t	0.074	0.052	0.053	0.048	0.037

It is evident that these data were notably segmented: for the "amplitude" runs, the enhancement of all five lengths of word populations were positive and hovered near statistical significance; in contrast, the "frequency" runs, although opted by the operators 44% more often, were all well within chance, and contributed little to the combined values. Detailed assessments of the 65 individual word contributions to these values revealed only two outlying components, the sequence 11000 being severely underpopulated ($t = -2.91$, $p_t < 0.004$), and likewise for 00011 ($t = -2.76$,

p_t < 0.006), but these occurred so rarely in the 512-bit sequences, and were sufficiently vulnerable to end-effects, that they should not be regarded as indicative without much larger databases.

Thus we were left with only an equivocal response to the hypothetical query that inspired the study, *i.e.*, a marginal restructuring of the binary word populations appeared, but only in the amplitude mode. The theoretical implications of this are far from clear, but nonetheless intriguing. Subsequently, this same drum output was successfully deployed in our Yantra experiment, providing a viable background environment that appeared to enhance the results.

C. Research Residue

We began this Section with a summary of an agenda of primary human/machine experiments that had established a substantial platform of consciousness-correlated phenomena that expressed itself in objectively quantifiable statistical deviations from chance behavior of various binary combinatorial systems, *e.g.* Benchmark REG, RMC, etc. These were followed by a litany of *ad hoc* explorations of specific correlational issues which to some extent reinforced the initial studies, but in other respects did not, displaying tendencies to replace the primary HI – LO effects with a myriad of secondary structural anomalies, some of which proved instructive in their own right, and a few of which were subsequently pursued more directly.

To complete the reportage of the full human/machine portion of our 28-year empirical program, we feel obliged to acknowledge a number of the other experimental projects that, for one reason or another, could not be refined to sufficiently stable operation to yield useful information. In addition to the "down-fountain" and hydraulic jump targets already mentioned, these included attempts to distort the optical fringe pattern of a Fabry-Perot interferometer; to create a temperature difference between two proximate electronic thermistors; to stabilize the bouncing of a small steel ball bearing on an oscillating glass plate; to delay or hasten the rate of phosphorescent decay of a photosensitive surface; to influence the patterns of gaseous discharge in a Crookes Tube apparatus; to alter the stress patterns in a photoelastic solid; to affect the binary information transfer along a microelectronic shift register; among others.

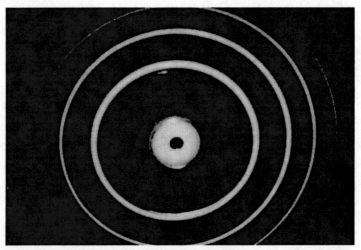

Fabry-Perot Interferometer fringe patterns

In most of these, although some provocative output deviations were observed, the calibration behavior of the equipment simply could not be stabilized sufficiently to provide a trustworthy basis for comparison with the proactive trials, even when differential protocols were utilized. Nonetheless, data files for all of these efforts have been retained, along with appropriate internal memoranda. In addition, a number of modest individual student or intern projects were conducted, some of which, like the ArtREG study described earlier, proved informative on specific issues.

References

[1] Lewis Carroll. *Through the Looking Glass: And What Alice Found There.* Philadelphia: Henry Altemus Company, 1897. p. 80.

[2] Frederick Reif. *Fundamentals of Statistical and Thermal Physics.* New York: McGraw-Hill, 1965. pp. 589–594.

[3] George Waddel Snedecor and William Gemmell Cochran. *Statistical Methods.* 6th edition. Ames: Iowa State University Press, 1967.

[4] David Acunzo, Samual Mas, and Baptiste Taillades. "Analysis and Interpretation of a Dataset from PEAR Laboratory." Report submitted to the Ecole Nationale Supérieure des Télécommunications de Bretagne, 2006. 6 pages.

[5] York H. Dobyns and Roger D. Nelson. "Interpretation of Drumbeat Statistics." February 21, 1992. 5 pages. PEAR internal white paper.

[6] York H. Dobyns and Roger D. Nelson. "Interpretation of Drumbeat Results." January 12, 1995. 13 pages. PEAR internal white paper.

12

SPACE/TIME INDEPENDENCE

*The Universal Mind has no time slit, no personal wall,
its knowledge is not limited by quantum probabilities.*
— Henry Margenau[1]

The remote perception portion of our consolidated PEAR program, to be reviewed in the next Section, well established that participants in such experiments (the "percipients") could acquire substantial objective and subjective information about physical targets far removed from their personal locations, where other individuals (the "agents") were situated at an agreed-upon time, without resort to normal sensory channels. More notably, it demonstrated that the accuracies of those anomalous perceptions were statistically independent of the degree of physical separation between the targets and the percipients, up to global distances. Even more remarkably, these experiments revealed that the perception efforts need not be performed at the same time as that of the physical target visitations; rather, the accuracies of the results were also statistically independent of temporal separations, up to several days, plus or minus. That is, information could be acquired about these targets *before* they were visited by the agent, or even before they were specified, which led to labeling this body of data "Precognitive Remote Perception," or "PRP."

Under the premise that the basic difference between the PRP anomalies and those of the human/machine experiments was that in the former the participant was *extracting* information from a random source (*i.e.*, the pool of potential targets and details thereof), whereas in the latter, information was being *inserted* into a random source (*i.e.*, the REG or other device outputs), it seemed reasonable to question whether similar spatial and temporal insensitivities might also characterize those anomalous human/machine interactions.

A carefully controlled program of remote/off-time REG experiments therefore was pursued, involving 30 operators and 491,000 trials per intention, arrayed in 265 series (including the remote data from the Benchmark RemREG experiments already described), with results strikingly similar to the laboratory-based data,[2] as well as to those of the PRP trials. Several extensive databases were acquired for which the operators and their target machines were separated by a wide range of distances up to several thousand miles. In the off-time variant, which comprised some 20% of the total database, the remote operators exerted their directional efforts up to several hours, or even several days, before, or after, the times of operation of the laboratory-sited machines.

As shown in the cumulative deviation plot of Figure II-27, the bar graph of Figure II-28, and the data summary of Table II-8, the total remote database acquired in this REG study compounded in a similarly anomalous fashion to that of the local experiments, with two noteworthy exceptions: 1) the effect size in the HI efforts was larger than that of the local Benchmark studies, and

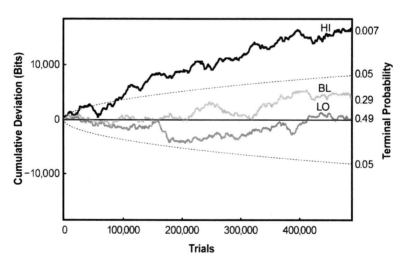

Fig. II-27. Cumulative deviations from theoretical mean, all remote REG data (265 series, 30 operators).

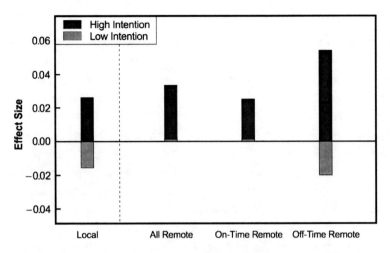

Fig. II-28. Comparisons of local, remote, and off-time REG effect sizes.

TABLE II-8

Remote REG Data Summary

Condition	Intention	Mean	z-score	Probability
On-Time Remote				
205 series,	BL	100.004	0.389	.348
391,000 trials	HI	100.027	2.348	.009
per intention,	LO	100.005	0.462	(.322)
29 operators	HI − LO	.022	1.333	.091
Off-Time Remote				
60 series,	BL	100.016	0.731	.232
100,000 trials	HI	100.054	2.416	.008
per intention,	LO	99.981	0.834	.202
10 operators	HI − LO	.073	2.298	.011
All Remote				
265 series,	BL	100.007	0.678	.249
491,000 trials	HI	100.032	3.185	7×10^{-4}
per intention,	LO	100.000	0.036	(.486)
30 operators	HI − LO	.032	2.227	.013

Note: probabilities in parentheses ( ) indicate effects opposite to intention.

2) that of the LO efforts was indistinguishable from chance. The former feature was mainly attributable to the *off-time* component performance; the latter, a hint of which asymmetry also appears in the local data, was primarily driven by the gender disparities to be described in the next chapter. In addition to manifesting the series position patterns mentioned in Chapter II-3 and described further in Chapter II-14, the trial count population distributions were similar to those of the corresponding local experiments at both the individual and collective operator levels, and the off-time segment of the total remote database displayed its proportionate share of these structural patterns.

Perhaps most instructive, however, was the evidence that the statistical scores of the remote data were essentially independent of the magnitudes of the intervening distances (Fig. II-29), thereby laying the axe to any conventional signal hypotheses wherein typical $1/r^2$ or exponential decays would have been expected. Similarly, those of the off-time data were essentially independent of the magnitudes of the intervening times (Fig. II-30), thereby invalidating any characteristic propagation velocity presumptions.

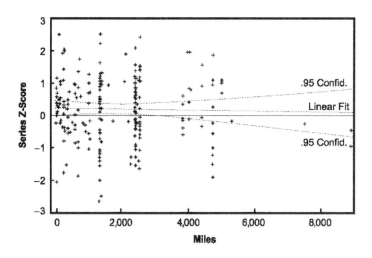

Fig. II-29. All REG remote HI – LO data vs. distance.

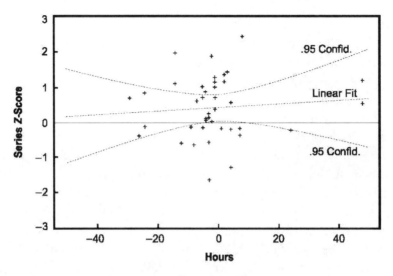

Fig. II-30. **All REG off-time remote HI – LO data vs. time difference.**

Given that similar remote/off-time effects were demonstrated on the Random Mechanical Cascade and Pendulum devices as well, this empirical removal of distance and time as salient correlates of both the human/machine and remote perception anomalies reinforced our initial suspicion that these two forms of anomaly entailed similar mechanisms of information exchange between human consciousness and random physical processes, albeit with opposite vectors of insertion and extraction. Beyond that, the absence of any statistically identifiable spatial or temporal attrition of the anomalous effects called into question the competence of any prevailing physical conceptualizations to encompass the phenomena. In fact, the demonstrably larger effect sizes of the off-time data, wherein the task becomes increasingly irrational and more fraught with uncertainty, forces consideration of much more radical dynamical propositions, which will be explored in Section IV.

In this context, we cannot avoid remarking on the continuing attention of certain theoretical physicists and philosophers of science to issues of "quantum entanglement," "separability," "nonlocality," and other conundrums of modern physics. While these commonly popularized topics are currently posed in somewhat different nomenclature and conceptual terminology, we suspect that the empirical demonstrations of the remote and precognitive capacities of the human mind to exchange objective and subjective information with widely separated physical systems and situations also may bear considerable relevance to these scientifically esoteric matters.[2]

Finally, we might point out that much of the criticism of experimentation in this field has focused on the inadequacy of shielding of the equipment from inadvertent or deliberate spurious disturbance by the operators, or by vibrational, acoustical, electromagnetic, chemical, or thermal means. For this reason, most laboratories take considerable care to preclude such artifacts via various noise suppression, vibration isolation, and electromagnetic shielding techniques, or by continuous observation of the operators. But the demonstration of equivalent patterns of results correlated with the intentions of operators who are thousands of miles away from the equipment and functioning at substantially different times, constitutes an even more comprehensive defense against such suspicions of artifactual disturbances, since by any reasonable criteria, these should be strongly dependent on the proximity of the operators. Any remaining suspicions of possible physical or procedural artifacts are thereby essentially disarmed by the double-blind, tri-polar protocols employed.[3]

References

[1] Henry Margenau. *The Miracle of Existence.* Woodbridge, CT: Ox Bow Press, 1984. p. 122.

[2] Brenda J. Dunne and Robert G. Jahn. "Experiments in remote human/ machine interaction." *Journal of Scientific Exploration, 6,* No. 4 (1992). pp. 311–332.

[3] Robert G. Jahn and Brenda J. Dunne. "Endophysical Models Based on Empirical Data." In R. Buccheri, A. Elitzur, M. Saniga, eds., *Endophysics, Time, Quantum and the Subjective: Proceedings of the ZiF Interdisciplinary Research Workshop, Bielefeld, Germany, 17–22 January 2005.* Singapore: World Scientific Publishing, 2005. pp. 81–102.

13

GENDER GAP

Breathes there a man with hide so tough,
Who says two sexes aren't enough?
— Samuel Hoffenstein[1]

The Co-Operator study described in Chapter II-4 yielded unexpectedly strong indications that the gender-pairings of the individual operators might be an important factor influencing their cooperative efforts. In fact, these results injected two possible implications into our broader search for correlates and phenomenological understanding: 1) subjective resonance may contribute to establishing propitious environments for anomalous human/ machine interactions; and/or 2) operator gender, *per se*, may be a relevant correlate of the phenomena. To pursue both of these possibilities, a comprehensive re-evaluation of several PEAR databases was undertaken to compare the performances of individual male and female operators over nine different experiments, the designs, protocols, and overall results of which have been described in previous chapters and detailed in the more comprehensive references,[2–8] *i.e.*:

- Local REG experiments
- Remote REG experiments
- Early PseudoREG experiments
- Local ATPseudo experiments
- Remote ATPseudo experiments
- Local RMC experiments
- Remote RMC experiments
- Local Pendulum experiments
- Remote Pendulum experiments

The conventional method for assessing male/female differences in performances in such protocols would be statistical comparisons of the respective distributions of results of the two genders for each experiment via z- or t-score calculations.[9] Given the substantial differences in the sizes of the 18 components of these composite databases, however, and the disparate array of measurables they entailed, a more straightforward approach was deployed to determine whether the gender disparities we had noted impressionistically could be substantiated. Specifically, we simply counted the numbers of male and female operators in each of the nine experiments who succeeded in producing results in their intended directions ($p < .50$), and the number exceeding chance expectations ($p < .05$), regardless of their individual effect sizes or the dimensions of their databases, and tabulated from these the proportions of operators in each group whose HI – LO, HI – BL, and LO – BL results correlated with their intentions. These overall results are summarized in Figure II-31 and Table II-9, with Table II-10 displaying separate summaries for each of the individual experiments. In the bar graph of Fig. II-31 z_{mf} denotes the statistical z-scores of the differences in male and female performances computed as described above.

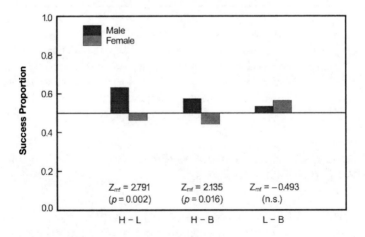

Fig. II-31. **Gender performance differences across nine experiments, pooled.**

TABLE II-9

Male vs. Female Operator Results of Nine Experiments

130 Male Operators			
	HI – LO	HI – BL	LO – BL
# Ops. $p < .50$	82	74	69
Proportion	.63	.57	.53
z-Score	2.98*	1.58	0.70
# Ops. $p < .05$	10 (3)	11 (1)	13 (4)
Proportion	.08 (.02)	.08 (.01)	.10 (.03)
z-Score	1.41 (−1.41)	1.81* (−2.21*)	2.62* (−1.01)
140 Female Operators			
# Ops. $p < .50$	64	61	79
Proportion	.46	.44	.56
z-Score	(−1.01)	(−1.52)	1.52
# Ops. $p < .05$	9 (5)	10 (8)	8 (5)
Proportion	.06 (.04)	.07 (.06)	.06 (.04)
z-score	0.78 (−0.78)	1.16 (0.39)	0.39 (−0.78)
Male/Female Differences			
$Z_{mf} p < .50$	2.79*	2.13*	−0.49
Probability	.003*	.016*	.311
$Z_{mf} p < .05$	0.33 (−0.33)	0.16 (−0.82)	0.66 (−0.16)
Probability	.37 (.37)	.44 (.21)	.25 (.44)

KEY

Ops. $p < .50$: Number of operators achieving in intended directions;

Proportion: Decimal proportion of number of operators achieving in intended directions;

z-score: z-score of number of operators achieving in intended direction;

Ops. $p < .05$: Number of operators achieving statistical significance in intended directions at <.05 level; * denotes significance of this number in intended direction; (*) denotes significance in opposite to intended direction.

Several salient features emerged from these results that may be helpful in illuminating the basic causes of the gender dichotomies:

- In general, the female operators tended to produce larger databases than the males. Across the nine experiments, the female databases constituted 69% of the data, compared to only 31% from male operators.

- Overall, the male data compounded to modestly significant mean shifts in the intended directions, achieved via reasonably balanced distributions, and well-behaved baselines.
- The female operators, on average, tended to produce larger (HI – LO) overall effects, but their data were strongly asymmetrical, with an average LO effect *opposite* to intention, and baselines that were systematically displaced toward the HI direction.
- Females tended to display larger variances than the males in their trial or run score distributions, an effect that manifested in their baselines as well as in their intentional efforts.
- The male/female differences in performance were much stronger and more distinct in the local experiments than in the remote, suggesting that the male/female disparities may be associated with the availability of visual feedback.

These gender-related patterns point to some underlying feature of the human/machine anomalies that most likely is related to the psychological, or possibly even physiological, characteristics of the human operators. These could include the nature of the information-processing mental dynamic that functions in such interactions; the psychological implications of "high," "low," and "baseline" intentions; and indeed, what is implied by the term "intention" itself. On this last point, our FieldREG experiments, described in an earlier chapter, showed that anomalous human/machine effects can be produced in the absence of any conscious intentions, or even of any conscious awareness, on the part of the operators or audiences.[10,11] These, along with the co-operator outcomes, re-assert that "intention," *per se*, may be only one factor stimulating the anomalies, and that the ability of an individual to establish a resonant bond with another participant, or with a machine, may be a factor of comparable, or even greater, consequence.[12]

TABLE II-10
Gender Differences in Each of Nine Human/Machine Experiments

Experiment	# Ops.	# Series	HI − LO δ	HI − LO %Ops. <.50	HI − BL δ	HI − BL %Ops. <.50	LO − BL δ	LO − BL %Ops. <.50
Local REG, male	50	228	3.2	66	0.4	58	−2.8	54
Local REG, female	41	294	4.7	34	1.8	37	−0.8	54
z_{mf}			3.04*		1.99*		0.00	
Remote REG, male	12	72	3.0	75	0.9	58	−2.1	58
Remote REG, female	15	140	3.4	53	0.6	47	−2.8	73
z_{mf}			1.13		−0.62		−0.77	
Local Pseudo, male	3	5	22.3	67	20.0	67	(7.8)	67
Local Pseudo, female	7	34	9.5	86	0.2	71	−8.9	71
z_{mf}			n/a		n/a		n/a	
Local ATPseudo, male	17	77	(−2.3)	47	(−3.5)	35	−2.0	59
Local ATPseudo, female	13	319	(−1.2)	38	1.9	46	(3.1)	46
z_{mf}			0.49		−0.60		0.71	
Remote ATPseudo, male	3	20	(−3.7)	63	(−1.4)	69	(2.3)	63
Remote ATPseudo, female	7	66	6.0	60	3.5	45	−2.5	70
z_{mf}			n/a		n/a		n/a	
Local RMC, male	16	40	0.38	63	0.23	69	−0.15	63
Local RMC, female	20	117	0.4	60	(−0.2)	45	−0.6	70
z_{mf}			0.18		1.43		0.42	
Remote RMC, male	3	6	0.4	67	0.6	67	(0.2)	00
Remote RMC, female	11	50	(−0.4)	45	(−0.4)	55	−0.03	55
z_{mf}			n/a		n/a		n/a	
Local Pendulum, male	20	61	0.011	60	.017	65	(.006)	50
Local Pendulum, female	20	121	(−.013)	30	(−.013)	35	−.001	40
z_{mf}			1.90*		1.90*		0.63	
Remote Pendulum, male	6	93	.010	83	.007	50	−.003	50
Remote Pendulum, female	6	32	.001	50	.000	50	−.000	50
z_{mf}			1.14		0.00		0.00	

Key

δ: Normalized deviation of the composite experimental mean from the theoretical expectation, multiplied by 100 for convenience of tabulation (vulnerable to statistical uncertainty for small data sets);

z_{mf}: Statistical z-scores of differences between male and female performances;

n/a denotes number of operators too small to calculate a valid z-score.

* indicates statistical significance ($p < .05$).

( ) indicates results opposite to intention.

Many of the female operators in these experiments also tended to produce distortions in their baseline data, which were ostensibly non-intentional conditions. In particular, the average female baseline was strongly displaced from chance, usually in the high direction, with unusually large variances in the output distributions, while the male baselines frequently displayed constricted variances that resulted in cumulative deviation traces that stayed much closer to the theoretical mean than would be expected. This evidence raises provocative questions about the generic reliability of "control" conditions in any scientific study, and suggests that the anomalous effects may be associated with some deeper level of consciousness, one perhaps more closely identified with the autonomic dimensions of the mind than with its cognitive ones.

If these gender-related disparities are somehow associated with biological processes, one is led to wonder what evolutionary benefit they may entail, and why the differences are more prominent in the local experiments than in the remote ones. To pursue this, we conducted a comprehensive review of the professional psychological literature on gender differences, which unfortunately is rife with inconsistent data and contradictory theories of its own.[13] The most persuasive and widely accepted findings seem to be associated with a female superiority in verbal ability, vis-à-vis a male advantage in visual-spatial skills, particularly in the capacity for "mental rotation," and in the characteristic distinctions in emotional responses. Superficially, none of these attributes appears to be directly relevant to our studies, but one might speculate that proficiency in spatial orientation would be of evolutionary benefit in establishing and defending territorial boundaries, and that verbal skills would provide an advantage in establishing interpersonal communications and group resonance. The graphic visual feedback in local experiments, which is not available in the remote experiments, could possibly stimulate some innate mode of information processing that favors the males in detecting some subtle spatial cues that aid them in producing better correlations with intentions and constricting their distribution variances, while the female capacity for enhanced communication may be reflected in their larger variances and effect sizes. In any event, it is curious that the opposite-sex co-operator results appeared to combine both the male propensity for directional symmetry and the female tendency to larger effect sizes.

It is important to recognize that while these surveys focused on the categorical distinction of biological gender, the variability in individual operator performances implies greater fundamental complexity of the phenomena. Since we were precluded from inquiring about the sexual preferences of our operators, our assessment was limited to the obvious physical categories. At one point a modest student project explored the performances of gay couples, pairing two such couples both by sexual preference and by biological gender as a control. The results of the bonded couples of same sex were comparable to those of our opposite-sex bonded couples, and the control pairs produced null results. Clearly, this study was too small to draw any substantive conclusions, but the results were suggestive. Any attempt to interpret these and other gender-related findings without taking into consideration this and many other relevant variables, such as individual information-processing strategies, sociological expectations, technological sophistication, personal belief systems, and a myriad of other potential cultural and environmental factors that might influence performance in tasks of this nature, not to mention possible physiological factors, will probably fall short of full understanding.

References

[1] Samuel Hoffenstein. *The Complete Poetry of Samuel Hoffenstein.* NY: Modern Library, 1954.

[2] Brenda J. Dunne. "Gender differences in human/machine anomalies. *Journal of Scientific Exploration, 12,* No. 1 (1998). pp. 3–55.

[3] Robert G. Jahn, Brenda J. Dunne, Roger D. Nelson, York H. Dobyns, and G. Johnston Bradish. "Correlations of random binary sequences with pre-stated operator intention: A review of a 12-year program." *Journal of Scientific Exploration, 11,* No. 3 (1997). pp. 345–367.

[4] Brenda J. Dunne, Roger D. Nelson, and Robert. G. Jahn. "Operator-related anomalies in a random mechanical cascade." *Journal of Scientific Exploration, 2,* No. 2 (1988). pp. 155–179.

[5] Brenda J. Dunne and Robert G. Jahn. "Experiments in remote human/machine interaction." *Journal of Scientific Exploration, 6,* No. 4 (1992). pp. 311–332.

[6] Roger D. Nelson, G. Johnston Bradish, Robert G. Jahn, and Brenda J. Dunne. "A linear pendulum experiment: Effects of operator intention on damping rate." *Journal of Scientific Exploration, 8,* No. 4 (1994). pp. 471–489.

[7] Robert G. Jahn, Brenda J. Dunne, and Roger D. Nelson. "Engineering anomalies research." *Journal of Scientific Exploration, 1,* No. 1 (1987). pp. 21–50.

[8] Roger D. Nelson, Robert G. Jahn, York H. Dobyns, and Brenda J. Dunne. "Contributions to variance in REG experiments: ANOVA models and specialized subsidiary analyses." *Journal of Scientific Exploration, 14,* No. 1 (2000). pp. 73–89.

[9] George E. P. Box, William Gordon Hunter, and J. Stuart Hunter. *Statistics for Experimenters: An Introduction to Design, Data Analysis, and Model Building.* Wiley Series in Probability and Mathematical Statistics. New York: John Wiley & Sons, 1978.

[10] Roger D. Nelson, G. Johnston Bradish, York H. Dobyns, Brenda J. Dunne, and Robert G. Jahn. "FieldREG anomalies in group situations." *Journal of Scientific Exploration, 10,* No. 1 (1996). pp. 111–141.

[11] Roger D. Nelson, Robert G. Jahn, Brenda J. Dunne, York H. Dobyns, and G. Johnston Bradish. "FieldREGII: Consciousness field effects: Replications and explorations." *Journal of Scientific Exploration, 12*, No. 3 (1998). pp. 425–454.

[12] Robert G. Jahn and Brenda J. Dunne. "Sensors, filters, and the Source of reality." *Journal of Scientific Exploration, 18*, No. 4 (2004). pp. 547–570.

[13] Brenda J. Dunne and Sepideh-Maria Ravahi. A Summary of Gender Differences in Human/Machine Anomalies Experiments and in the Psychological Literature. Internal Document.

14

NOW YOU SEE IT; NOW YOU DON'T

There is no such thing as a failed experiment,
only experiments with unexpected outcomes.
— Buckminster Fuller

Without contest, the most challenging aspect of consciousness-correlated anomalies experimentation is the well-known propensity of the phenomena to manifest with only irregular replicability. Beyond the already mentioned tendency of the sought-after effects to hide within the underlying random data sub-structures, other forms of longer-term irreproducibilies, wherein entire bodies of empirical data acquired using equipment, operator pools, protocols, and environmental conditions essentially identical to those of some effective previous study, often return substantially null results, or quite dissimilar albeit comparably anomalous effects to those of their predecessors. These capricious "hide-and-seek" characteristics have provided bountiful fodder for superficial skeptics who hail them as evidence of incompetent experimentation or delusional data interpretation. More profound contemplation, however, suggests that these apparent failures to replicate may be intrinsic features of the phenomena themselves and potentially valuable, if poorly understood, indicators of their fundamental nature. Given the prevalent reverence for replicability in contemporary science, this irregularity seems sufficiently hostile to the generic validity of the entire topic and to its ultimate comprehension to merit extensive definitive study. In this chapter we can mention only a few experimental and theoretical attempts to penetrate this mantle of irregular replicability that shrouds deduction of any causal chain that may be functioning in these situations.

A. Series Position

Perhaps the most commonly encountered form of this failure to replicate is the ubiquitous "decline effect," mentioned briefly in Chapter II-3, wherein initially promising anomalous results, when pursued into second- and third-generation experiments of similar format, gradually erode into insignificance, often leading to frustrated abandonment of the study by the investigators, and to consequent guffawing by the skeptics. It is evident from the outset that any rigorous pursuit of such a temporal evaporation of the anomalous effects must necessarily labor under a stringent caveat to obtain huge individual and collective datasets if definitive patterns are to be established.

Reference 1 summarizes a major portion of such an exhaustive study, which in fact yielded some enlightening results. Namely, the directional effect sizes achieved by the operators in a broad range of human/machine experiments showed well-defined patterns of correlation with the ordinal positions of the experimental series in both the collective and individual databases. Specifically, while there were indeed statistically significant tendencies for operators to produce better scores in their first series, then to fall off in performance in their next few series, these were then found to recover to some intermediate levels over subsequent series, eventually stabilizing to characteristic asymptotic values (Figure II-32). Such patterns of ordinal correlations appeared in both the local and remote versions of extended sequences of several of the experimental protocols described in previous and subsequent chapters, *with no such patterns appearing in the baseline or calibration data*. In short, there were indeed decline effects, but they usually presented only as initial phases of more complex patterns of performance evolution.

These persistent patterns of serial decline and recovery were too prominent to be ignored, yet too enigmatic and complex to support any obvious simple interpretation. The absence of such tendencies from the baseline data or machine calibrations, and their ubiquitous occurrence over many different types of devices

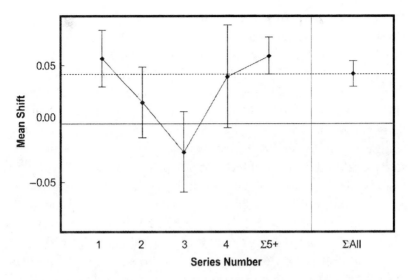

Fig. II-32. Overall HI – LO series position patterns obtained by
"prolific" operators in Benchmark experiments.[1]

and protocols, strongly suggested that these trends reflected some
characteristics of the operators and/or the experimenters, rather
than being artifacts of machine performance. (It may be relevant
to note in passing that these serial patterns bear some resemblance to the damped cosinusoidal "switching transients" that
characterize the onsets and interruptions of various forms of mechanical and electromagnetic systems, both physical and biological, as they eventually converge onto new levels of operation.)

Reference 1 includes an extensive Appendix that surveys the
psychological literature dealing with serial position effects associated with learning, memory, perception, and other "common"
cognitive processes. The broad similarities of the patterns in our
data to those reported therein under nomina of "decline," "primacy," "recency," and "terminal" effects propounded in that literature
further supports the suspicion that some fundamental attribute of
the human psyche might be manifesting here. At the least, it appears that some types of *subjective* factors are participating in the
generation of these *objective* anomalous effects.

B. Inter-Laboratory Replications and Structural Anomalies

In Chapter II-10 we described one bewildering inter-laboratory replication effort with a double-slit device that returned substantially dissimilar results despite using the same basic technology and procedures in two different laboratories. Another empirical demonstration of an even more complex genre of reproducibility confounds appeared in the context of a major REG replication study undertaken in a collaboration of our laboratory with the Institut für Grenzgebiete der Psychologie und Psychohygiene at Freiburg and the Justus-Liebig-Universität at Giessen. Using similar equipment and protocols, the three laboratories performed a long and detailed agenda of REG experiments, the results of which are fully recorded in Reference 2. To summarize, whereas overall HI – LO mean separations, which were the pre-specified primary criterion of this replication effort, compounded in the intended directions at all three laboratories, the sizes of these failed to attain those of the original PEAR benchmark experiments, or even to achieve persuasive levels of statistical significance on their own. However, pre-planned analyses of a number of secondary parameters maintained in this study revealed that several interior structural anomalies also departed from chance. For example, utilizing a Monte Carlo simulation technique* that precluded any multiple testing artifacts, it was possible to demonstrate that the observed assortment of departures from the null hypothesis expectations in the individual and collective datasets were highly significant. It was as if the simple displacements of the mean that had characterized the Benchmark experiments had been partially transformed into a number of more subtle anomalous fragments in the new data. These included such features as a reduction in trial-level standard deviations, distortion of series-position patterns,

---

* Monte Carlo techniques refer to making many runs of a model or simulation using random values different in each run for all unknown or uncontrollable parameters so as to obtain a sound statistical estimate of the model's behavior.[3]

correlations with operators' prior experience, and differential dependencies on various secondary parameters such as feedback type or experimental run length, that were separately and collectively well beyond chance expectations.

Prompted by these complicating findings, several retrospective analyses of other existing PEAR databases were undertaken in a search for similar distortions, and additional replication experiments were initiated to test this transformation hypothesis. From these it became progressively more apparent that on some occasions such interior structural anomalies could indeed substitute for, or simply supplement, the primary mean-shift effects intended by the operators. On other occasions, however, the original modes of anomalous expression reappeared as before. The criteria for such metamorphoses were far from evident.

C. Theoretical Implications

In an attempt to tackle the replication issue at a more epistemological level, we also collaborated with certain theoretical physicists on conceptual models which proposed treating mind/matter interactions as generalized complex systems, for which linear first-order approaches were known to be both phenomenologically and methodologically inadequate and required more sophisticated non-linear second-order techniques.[4,5] In the course of these studies, we came to believe that only in these frameworks could the reproducibility issue be properly posed and analyzed, and the inclusion of subjective factors in the dynamical formation of models of these and other consciousness-related phenomena be comprehensively attempted.

D. Generalizations, Associations, and Speculations

Further pursuit of the replicability issue should probably acknowledge the multitude of more common experiential phenomena that are routinely accepted as intrinsically idiosyncratic and unpredictable. In the subjective domain, for example, we find great variations in individual and collective responses to artistic stimuli, opportunities, disappointments, challenges, demands, threats, etc. It is a common occurrence that an attempt to create a sequel to a brilliant work of art frequently falls far short of its predecessor, and in the lore of the theater it is anticipated that the second performance of a production that enjoyed an acclaimed premiere may turn out to be mediocre. Our biological responses to physical stimuli also vary greatly from one exposure to the next, as well as from one person to another, and major "objective" physical phenomena, such as volcanic eruptions, tsunamis, earthquakes, or meteor impacts, are by their very nature essentially irreplicable and unpredictable. For some of these, the common denominator is an over-riding investment of human or non-human subjective consciousness; others can be rationalized as having patterns of precipitating causation too complex to model, let alone to replicate. But there may be yet a deeper layer of origin to be considered; *i.e.* in addition to the well-accepted function of consciousness to recognize patterns, it may well have some capacity to *create* them.

As we shall discuss in more detail in Section V, over the latter portion of his career, the renowned psychoanalyst Carl Jung proposed a genre of phenomena he labeled "acausal" — effects which displayed no deterministic source, *i.e.*, which could not be traced back to prior causative conditions. They simply happened, unpredictably.[6] Clearly, if such indeed exist, they certainly cannot be constrained by any replicability criterion of their own. Rather, they constitute the ultimate array of random events, and if our laboratory-based effects share any common features with them, some violations of standard scientific replicability rules are to be expected.

All of this is not to claim any unity of understanding of the irreplicability of the anomalous effects we have studied, or of their more broadly cast common cousins, but only to point out that they have much equally inexplicable company in the broader context of human experience. It is not unreasonable to speculate that phenomena that are intrinsically associated with the processes of human consciousness will inevitably reflect influences such as anticipation, fear, novelty, playfulness, past experience, expectations, desires, and any other factors that entail greater or lesser degrees of uncertainty, and thus do not lend themselves readily to deterministic causality. In a similarly speculative tone, might we not also suspect that any phenomenon that entails such characteristically complex randomness at its microscopic level may also display a comparably chaotic replicability in its collective manifestations?

References

[1] Brenda J. Dunne, York H. Dobyns, Robert G. Jahn, and Roger D. Nelson. "Series position effects in random event generator experiments." *Journal of Scientific Exploration, 8*, No. 2 (1994). pp. 197–215.

[2] Robert Jahn, Brenda Dunne, G. Johnston Bradish, York Dobyns, Arnold Lettieri, Roger Nelson, Johannes Mischo, Emil Boller, Holger Bösch, Dieter Vaitl, Joop Houtkooper, and B. Walter. "Mind/Machine Interaction Consortium: PortREG replication experiments." *Journal of Scientific Exploration, 14*, No. 6 (2000). pp. 499–555.

[3] Nicholas Constantine Metropolis and Stanislaw Ulam. "The Monte Carlo method." *Journal of the American Statistical Association, 44* (1949). pp. 335–341.

[4] Harald Atmanspacher and Robert G. Jahn. "Problems of reproducibility in complex mind-matter systems." *Journal of Scientific Exploration, 17*, No. 2 (2003). pp. 243–270.

[5] Vasileios Basios. "Complexity, Interdependence, and Objectification." In Zachary Jones, Brenda Dunne, Elissa Hoeger, and Robert Jahn, eds., *Filters and Reflections: Perspectives on Reality*. Princeton: ICRL Press, 2009. pp. 187–207.

[6] Carl Gustav Jung. *Synchronicity; an Acausal Connecting Principle.* Trans. Richard Francis Carrington Hull. 2nd ed. Princeton, NJ: Princeton University Press, 1973. 135 pp. Bollingen series no. 20.

15

INCONCLUSIVE CONCLUSIONS

All knowledge degenerates into probability.
— David Hume[1]

The mosaic of empirical human/machine results presented in the preceding chapters and their supporting archival references is far too complex and ambiguous to be cleanly summarized in any one definitive format. For an academic research program, the PEAR laboratory entailed an unusually extensive and diverse complex of participants, projects, equipment, protocols, databases, analytical techniques, theoretical models, and alternative interpretations. At best, we can only reflect on the large number and broad variety of experimental enterprises, the huge bodies of results accumulated, and the many common operators who contributed to them. Any temptations to agglomerate all of these into composite figures of merit should be dismissed, for a number of practical and philosophical reasons. First, the panoply of studies spanned such a diversity of measurables, protocols, database sizes, statistical methods, technologies, and subjective factors that it is unlikely that any omnibus criteria that would satisfy all analyses could be posed. Second, such "bottom line" argumentation would inevitably obscure much of the secondary structural anomalies that spontaneously appeared in many of the empirical datasets. Finally, and perhaps most pertinently, the great majority of the experiments actually were not directed to achievement of extra-chance results, *per se*; those had been more than adequately demonstrated in the early years of the program, certainly by the completion of the extensive 12-year Benchmark era.[2] At this point, it was in fact internally and externally proclaimed that the bulk of the experiments going forward would henceforth rather be directed toward specific characteristics and parametric correlations of the effects that might aid in fundamental understanding and theoretical modeling of the phenomena, at some potential expense to the scale of the anomalous data deviations. Given this attitude, it now seems more scientifically astute and responsible simply to concede the presence and complexity of some overarching anomalous mechanism, and to advocate continuation of a creative, albeit disciplined scholarly search for its illumination.

In so attempting, we might reflect that in science, as in most human affairs, we instinctively invoke the primordial dialogue of experience with deduction, experiment with theory, manifestation with model, in lifelong efforts to progress toward understanding of the mechanics of life and of life's physical environments. In the modern "exact" sciences, as in most matters of contemporary business, law, medicine, and technology, we have disciplined ourselves to restrict such dialogues to deal in objective information ("hard facts"); causal temporal chains ("if this, then that"); and demonstrably predictable behaviors ("laws" of action and reaction). This conventional, albeit restrictive paradigm of logic has the advantage of relatively unambiguous conceptualization, specification, and quantification, and indisputably has led, and will continue to lead, to well-marked progress and coherence of interpretation in those domains to which it is applicable.

Unfortunately (or perhaps fortunately), not all human experience plays by these rules and submits to these presumptions. Everyday life is replete with many forms of uncertainty and the events and reactions that derive therefrom correspondingly defy causal determinism, yet are evidentially indisputable. Our laboratory presence and its work have steadily attracted personal and anecdotal reports from many rational and cogent people who have experienced inexplicable malfunctions of technical equipment, precognition of subsequently confirmed future events, or various types of non-local communication. In such situations the chains of causality become much more complex, and in most cases totally obscure. The extensive array of more rigorously acquired, but collectively contradictory research data summarized in this Section adds its own more disciplined form of testimony to the reality of such an irregular domain of experience, and leaves us intellectually mired in an "acausal" swamp. It perhaps may be useful here to summarize the several categories of departure of our empirical observations from conventional deterministic behavior and practice.

A. Sub-Structural Scenarios

Although the latter-day appearances of sub-structural idiosyncrasies both in the human/machine studies just reviewed and in the remote perception categories to follow, clearly complicate the identification and statistical specification of genuinely anomalous responses, they also may provide important hints on some fundamental aspects of the phenomena. First of all, we should recognize that such gross irreplicability is by no means limited to these types of interactions; it can readily be observed in many common physical events. For example, if one were to shatter, one by one, an ostensibly identical sequence of dinner plates by firing identical bullets from the same identically positioned rifle, it is doubtful that the detailed patterns of breakage would be anything like identical. Alternatively, if one were to take a sequence of research snapshots of the turbulence forming in fluid flows around solid obstacles, no two patterns would be identical. The reasons: the microscopic uncertainties underlying either of these processes would be sufficient to escalate into macroscopic patterns that are essentially unpredictable. We suggest that similar mechanics of microscopic uncertainties predicating macroscopic variabilities may be prevailing in these consciousness-related anomalies experiments, as well.

In all of these situations, we suspect some correspondence to the non-linear dynamical behavior of so-called complex and chaotic systems, wherein their initial conditions and physical constraints cannot be specified to sufficient accuracy to guarantee any particular paths of subsequent behavior. Rather, a plethora of possibilities can be observed that can be described only in statistically generalized terms, and in suspiciously subjective language (*e.g.* "strange attractors").

The overarching message here is an emerging escalation of the importance of *uncertainty* within the panoply of consciousness-correlated anomalous causes and effects. Whether this uncertainty prevails at a quantum or macroscopic physical level seems less important than that it can compound and predicate observable deviations from rigid predictability. More directly pertinent is the tantalizing possibility that even some form of subjective uncertainty in the mind of the human participant can also prompt the complex dynamic, as suggested by the "serial position" patterns of achievement described in Chapters II-4 and II-14. Specifically, initial exposure to an experiment inherently involves personal uncertainties of performance; success therein imparts some confidence; which reduces the uncertainty during subsequent attempts, thereby predicating less success; which partially re-introduces the uncertainty; and the cycle repeats showing the characteristic oscillations of performance actually observed, until stabilized by some balancing of these conflicting operator impressions. An even more provocative interpretation has been suggested by several other bodies of research indicating that similar decline effects appear in numerous other scientific contexts, notably including medicine[3] and biology.[4] These cast doubt on the universal validity of the conventional concepts of replicability and control study techniques, and further open the door to the intrusion of subjective factors that can introduce wider uncertainties into statistical patterns of results.[5,6,7]

Likewise, the sub-structural patterns of achievement that were originally found in the three-laboratory replication study described in the previous chapter, and subsequently uncovered in several other studies via Monte Carlo simulation techniques, also bear similarities to the non-linear behaviors of the complex and chaotic systems just mentioned. In this case, ostensibly similar experimental conditions of protocols, participants, equipment, and environments, produced inexplicably different patterns of results than the predecessor studies they were endeavoring to replicate. Yet, their composite patterns were every bit as anomalous in their own ways as the originals.

Our considered interpretation: given the surfeit of technical, operational, personal, and cultural uncertainties rampant in such studies, the consciousness influences may alternatively express themselves in an orderly deviation from chance behavior, as in the initial studies; in erratic patterns of anomalous sub-structures of output data, as in the serial position effects or the replication studies data; or in a myriad of unpredictable combinations of both modalities. Einstein is reported to have opined: "God does not play dice." Perhaps not, but it appears that He may play "pickup sticks," or "Texas hold 'em," or even more complex games involving self-reflexive creative organization of random sources.

B. Non-Local Logic

The stark absence of common physical properties, most notably distance, time, and the particular kind of random source involved, from the list of empirical correlates observed in these experiments gives clear warning that we are not dealing here with any form of conventionally disposed information transfer process. Rather, as Larry Dossey has well termed it, this is strong indication of a "non-local" capacity of the human mind,[8] the implications of which far exceed any specific explications of our particular experiments. If indeed the effects studied here can manifest full strength at global distances and with substantial time displacements, the underlying information mechanics must fall far outside of any process currently recognized by conventional science. While this feature clearly complicates the equipment protection issues raised in Chapter II-4, on the positive side it also opens vastly wider opportunities for beneficial applications, if the phenomenological basis can better be comprehended and the requisite technology brought to heel.

C. Physicality and Randomicity

But even in this imposing non-local, atemporal framework, another critical physical criterion remains to be specified, namely whether totally random sources or processes are essential to establishing the requisite entanglement with the participating consciousnesses. As discussed in Chapter II-6, we have so far failed to resolve this issue empirically, possibly because it is inherently precluded from resolution by a hierarchy of cascading confounds. For example, even assuming that a particular device or process can be constructed that qualifies as totally pseudorandom (*i.e.* deterministic), what about the attitude of the operator in addressing the task, or in reacting to the technical or emotional, local or non-local environment? Could not these subjective factors affect, say, the time or place of incursion into the (finite length) pseudorandom bit string, or could they even have affected its original conception and preparation? As Sherlock Holmes was fond of saying to Dr. Watson: "These are very deep waters."[9] The point is, that traced far enough back on the chain of causation, ultimately some random element must be encountered. Perhaps that is sufficient to render *any* device or process potentially vulnerable to its interacting consciousnesses.

D. Scale Issues

Many credible reports can be found in the historical and contemporary literature regarding scholarly assessment of *macroscopic-scale* anomalous physical phenomena: poltergeists, levitations, hauntings, survivals and reincarnations, translocations, theological miracles of many cultures, etc., wherein the apparent scales of forces and energies manifested and manipulated are purported to attain major dimensions. In the course of our own studies, we have retained some scholarly cognizance of these imposing special classes of anomalous experience, under the suspicion that they may offer valuable hints about the taxonomy of the generic phenomena that are pertinent across all scales and contexts. But because of the unpredictability of their occurrence and the corresponding difficulties of performing well-controlled empirical research, and given our own charter of commitment to issues of more common engineering relevance, we have preferred to focus primarily on microscopic-scale effects. While this strategy has yielded a somewhat more amenable degree of statistical tractability, not surprisingly it has largely constrained the evidence to a microscopic scale as well, thus requiring very sensitive equipment and very large databases to demonstrate and correlate it convincingly. In this respect, the scenario has not been unlike that prevailing at the dawn of the quantum age, when atomic-scale departures from classical expectations became evident only with the advent of optical and electronic equipment of adequate sensitivity to extract them from their backgrounds of classical random noise. But in the foreground of potential future applications lurked egregiously macroscopic phenomena.

The consistent appearance in our small-scale studies of anomalous signal-to-noise ratios of the order of 1 part in 10^4 across numerous devices, operators, and types of experiment, while clearly challenging to the acquisition of a credible inventory of effects, nonetheless is a potentially valuable indication of the fundamental mechanism of their establishment. Although derided even in some parapsychological research quarters as an uninstructive focus on "itsy-bitsy bias" in contrast to dogged searches for larger effect sizes from "gifted" subjects, within the mission of technological relevance we have defined for our work, our ability to study this range of effects has greatly enhanced the operator pools, and more importantly, the public relevance of the issue to much broader constituencies. In fact, it appears that this scale of anomalous experience is much more routinely available to a large portion of human culture, and to a wide range of technical, social, and personal applications.

E. Metaphoric Mechanics

Section IV of this book will present a few of our theoretical attempts to grasp the essence of the strange phenomena we have been privileged to study over the past three decades. At best, the models proposed, much like virtually all physical science theories, are metaphoric in character; essentially they say: "Gee, this reminds me of another experience I have had in another context, whose validity I have come to accept in that format, and which seems to submit to a similar conceptual dynamic..." These concessions to metaphor do not necessarily diminish the models. If they aid in conceptualization, guide further experimentation and the interpretation thereof, and open opportunities for useful applications, they may be worth retaining in the analytical workshop for whatever practical uses they may offer. But none of these descriptive analogies should be proffered as a comprehensive definition of the deeper source of the phenomena. And in this particular case, we are a very long way from any such grand epistemological achievement, if indeed such ever can be attained.

Rather, using a less ambitious but more pragmatic approach, we can lay the various proposed metaphoric models alongside one another in a search for similar conceptual features that appear in each of them, in the hope that these common denominators may help to triangulate certain aspects of their common Source. Not surprisingly, indeed somewhat reassuringly, the list of these primarily subjective features is essentially identical to that inferred from the experimental experiences of our operators, *e.g.* intention, resonance, openness, unconscious processing, gender, and physical or emotional uncertainty. While these subjective prevalences do not guarantee appearance of the anomalous effects, they frequently seem to be fellow travelers with them, to the extent that we might regard them as environmental conditioning agents, pending eventual identification of any more explicit biological, psychological, or physical chains of causation, if in fact such exist.

Empirically, all of these anomalous consciousness-related phenomena, spanning many venues of physical manifestation, waxing, waning, and changing form elusively from day to day, person to person, and site to site, have been found to correlate far more with inherently subjective parameters than with nicely specifiable objective dimensions, inexorably forcing us to confront a very fundamental philosophical dilemma: Can science as we know it constructively accommodate such renegade phenomena within its scholarly purview, or must it accept circumscription short of this bizarre territory? This will be the overarching challenge of our theoretical explorations to follow, and of our exhortations to broaden the scientific "rules" to encompass subjective factors that are currently excluded.

But first we must update the salient results of another potentially indicative body of experimental and analytical research on a form of patterned clairvoyance we have termed "remote perception."

References

---

[1] David Hume. *A Treatise on Human Nature* (1739–40), ed. L. A. Selby-Bigge (1888), book 1, part 4, section 1. p. 180.

[2] Robert G. Jahn, Brenda J. Dunne, Roger D. Nelson, York H. Dobyns, and G. Johnston Bradish. "Correlations of random binary sequences with pre-stated operator intention: A review of a 12-year program." *Journal of Scientific Exploration, 11*, No. 3 (1997). pp. 345–367.

[3] John P. A. Ioannidis. "Contradicted and initially stronger effects in highly cited clinical research." *The Journal of the American Medical Association, 294*, No. 2 (2005). pp. 218–228.

[4] Michael D. Jennions and Anders Pape Møller. "Relationships fade with time: A meta-analysis of temporal trends in publication in ecology and evolution." *Proceedings: Biological Sciences, 269*, No. 1486 (2002). pp. 43–48.

[5] Jonah Lehrer. "The Truth Wears Off." *The New Yorker,* December 13, 2010. pp. 52–57.

[6] Jonathan W. Schooler. "Introspecting in the spirit of William James: Comment on Fox et al. (2010)." *Psychological Bulletin* (in press). footnote.

[7] Dick J. Bierman. "On the nature of anomalous phenomena: Another reality between the world of subjective consciousness and the objective work of physics?" In Philip R. van Loocke, ed., *The Physical Nature of Consciousness (Advances in Consciousness Research 29)*. New York: John Benjamins, 2001. pp. 269–292.

[8] Larry Dossey. "PEAR Lab and Nonlocal Mind: Why They Matter." Foreword to *EXPLORE: The Journal of Science and Healing, 3*, No. 3 (May/June 2007). pp. 191–196.

[9] Sir Arthur Conan Doyle. "The Adventure of the Speckled Band." In *The Complete Sherlock Holmes*. New York: Doubleday, 1930. p. 263.

SECTION III

Remote Perception:
Information and Uncertainty

"Oracle of the Pearl"

1

SECOND SIGHT

...Celestial light,
Shine inward, and the mind through all her powers
Irradiate; there plant eyes; all mist from thence
Purge and disperse, that I may see and tell
Of things invisible to mortal sight.
— John Milton[1]

While the human/machine studies at PEAR were exploring the possibility that consciousness could *introduce* information into its physical environment, another major portion of the program was investigating the phenomenon of remote perception: experiments that suggested that consciousness could also *extract* information from its environment independently of distance, time, or the usual sensory processes. Dubious as such a proposition might appear in the context of 21st-century science with its unyielding emphasis on deterministic materialism, this capacity has been acknowledged in virtually every culture since the dawn of human civilization, under a multitude of names, such as divination, prophecy, scrying, clairvoyance, or second sight.

In the 16th century, the renowned alchemist, physician, and philosopher, Philippus Theophrastus Aureolus Bombastus von Hohenheim, also known as "Paracelsus," declared in a section of his writings devoted to the role of "active imagination" in human representation of the universe:

> Man also possesses a power by which he may see his friends
> and the circumstances by which they are surrounded, al-
> though such persons may be a thousand miles away from
> him at that time.[2]

A century later, Emanuel Swedenborg, whose staggering scientific and engineering insights in fields as diverse as physics, astronomy, geology, and chemistry, included prescient sketches of inventions as varied as a flying machine, a submarine, and a fire engine, was also a deeply mystical scholar who testified to frequent experiences of clairvoyance. In one commonly cited incident, while dining with friends in Göthenburg he accurately described a fire that was raging near his home in Stockholm, 300 miles away. This perception was witnessed by several observers, and later investigated and validated by Immanuel Kant.[3]

In the more recent history of Western science, a considerable body of literature describing scholarly investigations of "extrasensory perception" already had been amassed when, in the mid-1970s, a new scientific protocol for empirical investigation of the phenomenon was introduced by physicists at Stanford Research Institute (SRI) under the name "remote viewing." Although these studies were conducted under the auspices of a classified government program, the researchers published brief descriptions of their findings in the mainstream scientific journals *Nature*[4] and the *Proceedings of the IEEE*.[5] These articles included several examples of participants' descriptions of remote geographical locations, inaccessible to them by any usual sensory means, which were virtually photographic in accuracy, analysis of which yielded overall statistical results that were well beyond chance expectations.

Intrigued by these reports, particularly by their almost casual reference to the ability of some of their participants to acquire their anomalous information *before* the targets had actually been selected, one of us (B.D.) conducted a modest replication in the Chicago area of the SRI experiments. This followed the same basic protocol, but used inexperienced volunteers, rather than people who claimed exceptional abilities.[6] Impressionistically and statistically, the results of most of these trials also were highly significant and frequently displayed remarkable correlations between the perceptions and the actual details of the target sites. Further studies explored the effects of increasingly greater spatial and temporal separations and of more than one person addressing the same task.[7] The encouraging results of those studies provided the impetus for what eventually became one of the two principal components of the PEAR experimental program.

Much of this prior work has been presented elsewhere—in our earlier book,[8] and in several archival publications[9,10] and technical reports[11-13]—wherein have been displayed several examples of the striking correspondences between remote physical target scenes and the perceptions thereof obtained by untutored human percipients, along with descriptions of some of the scoring methods developed at PEAR for quantitative evaluation of the anomalous information acquired by these empirical techniques. In this Section, we shall attempt to summarize the evolution of these experiments and analyses, and focus on their convergence on enlightening realizations regarding the character of the underlying phenomenon.

References

[1] John Milton. *Paradise Lost: The Third Book*. The Harvard Classics, 1909–14.

[2] Franz Hartmann. *Paracelsus: Life and Prophecies*. Blauvelt, NY: Rudolf Steiner, 1973. p. 105.

[3] Immanuel Kant. *Dreams of a Spirit-Seer, and Other Related Writings*. Trans. Emanuel F. Goerwittz. Ed. Frank Sewall. New York: MacMillan, 1900.

[4] Russell Targ and Harold E. Puthoff. "Information transmission under conditions of sensory shielding." *Nature, 251*, No. 5476 (1974). pp. 602–607.

[5] Harold E. Puthoff and Russell Targ. "A perceptual channel for information transfer over kilometer distances: Historical perspective and recent research." *Proceedings of the IEEE, 64,* No. 3 (1976). pp. 329–354.

[6] Brenda J. Dunne and John P. Bisaha. "Precognitive remote viewing in the Chicago area: A replication of the Stanford experiment." *Journal of Parapsychology, 43* (1979). pp. 17–30.

[7] John P. Bisaha and Brenda J. Dunne. "Multiple subject and long-distance precognitive remote viewing of geographical locations." In Charles T. Tart, Harold E. Puthoff, and Russell Targ, eds. *Mind at Large: IEEE Symposia on the Nature of Extrasensory Perception.* New York: Praeger Special Studies, 1979. pp. 109–124. (Reprinted: Charlottesville, VA: Hampton Roads Publishing Company, 2002. pp. 98–111.)

[8] Robert G. Jahn and Brenda J. Dunne. *Margins of Reality: The Role of Consciousness in the Physical World.* San Diego, CA: Harcourt Brace Jovanovich, 1987. Reprinted: Princeton, NJ: ICRL Press, 2009. Section III, pp. 149–191

[9] Brenda J. Dunne and Robert G. Jahn. "Information and uncertainty: 25 years of remote perception research." *Journal of Scientific Exploration, 17,* No. 2 (2003). pp. 207–241.

[10] Robert G. Jahn, Brenda J. Dunne, and Eric G. Jahn. "Analytical judging procedure for remote perception experiments." *Journal of Parapsychology, 44* (1980). pp. 207–231.

[11] Brenda J. Dunne, Robert G. Jahn, and Roger D. Nelson. "Precognitive Remote Perception." Technical note PEAR 83003. Princeton Engineering Anomalies Research, Princeton University, School of Engineering/ Applied Science, Princeton, NJ. August 1983.

[12] Robert G. Jahn, Brenda J. Dunne, Roger D. Nelson, Eric G. Jahn, Todd Aaron Curtis, and Ian A. Cook. "Analytical Judging Procedure for Remote Perception Experiments — II: Ternary Coding and Generalized Descriptors." Technical note PEAR 82002. Princeton Engineering Anomalies Research, Princeton University, School of Engineering/ Applied Science, Princeton, NJ. 1982.

[13] Brenda J. Dunne, York H. Dobyns, and Susan M. Intner. "Precognitive Remote Perception — III: Complete Binary Data Base with Analytical Refinements." Technical note PEAR 89002. Princeton Engineering Anomalies Research, Princeton University, School of Engineering/ Applied Science, Princeton, NJ. August 1989.

2

PRP AT PEAR

*We know from the experimental data of psi research
that [a] viewer in the laboratory can focus his or her
attention anywhere on the planet and, about
two-thirds of the time, describe what is there.*
— Russell Targ[1]

This second major empirical program at PEAR came to be known
as "Precognitive Remote Perception," or PRP, rather than using
the earlier nomenclature of "remote viewing." Most of the experimental trials were carried out in a precognitive mode wherein
the remote target scenes were described by individuals called
percipients, before those locations had been specified, much less
visited by the outbound participants, or *agents*, and many of these
percipients maintained that their experiences were not, strictly
speaking, of a simple visual nature, but entailed other categories
of impression. A substantial number of trials also were executed
retrocognitively, wherein perceptions were generated well after
the agents had visited the targets, and a smaller number were
performed in "real time."

In its basic form, the PEAR PRP experimental protocol required the percipient, without access to any conventional sensory
input, to attempt to sense and describe the physical and emotional
aspects of a randomly selected geographical site at which an agent
would be stationed at a specified time. Percipients were asked to
spend 15 to 20 minutes attempting to visualize or experience the
target scene and to record these impressions in a free-response,
stream-of-consciousness form, either orally into a tape recorder
or in writing, optionally including drawings. The agents, who in
almost all cases were known to the percipients, were instructed

to situate themselves at the target sites at the agreed-upon times and to immerse themselves subjectively and objectively in the scenes for about 15 minutes. At the close of the visitation periods, they were asked to record their impressions of the target scenes, and when possible, to supplement them with photographs and/or drawings to corroborate their verbal descriptions.

Unlike some of the procedures followed in remote viewing studies elsewhere, where percipients were trained to use particular techniques, or where perceptions were generated in a laboratory setting with an experimenter present and actively eliciting information, PEAR percipients were free to select their own preferred locations and their own personal strategies for addressing the task, and experimenters were not present during the perception efforts. The agents, like the percipients, were also free to employ their own subjective approaches to the task, and were simply encouraged to attempt in some way to share their experiences of the targets with the percipients. Although participants were given no further instructions, an attitude of playfulness was encouraged with emphasis placed on enjoyment of the experience rather than on the achievement *per se*. This permissive stance notwithstanding, in all cases strict precautions were taken to ensure that perceptions were recorded and filed before percipients had any sensory access to information about the targets, and no ordinary means of communication between percipients and agents were permitted until the trials were completed and recorded.

It was understandable, therefore, that the transcript styles of individual percipients varied widely, ranging from a few cryptic details to lengthy impressionistic flows of imagery. No records were maintained on the relative effectiveness of the various personal strategies deployed by the participants or on any of their psychological or physiological characteristics. They were encouraged, however, to provide informal accounts of their experiences, and these anecdotal descriptions provided useful glimpses into some of the more qualitative aspects of the underlying process. For example, some percipients commented that they found it helpful to clear their minds, visualize a blank screen, and wait for an image of the agent to appear. Others indicated that they preferred to address the task with repeated brief intervals of attention, rather than with sustained focus. Some agents reported that they would imagine the percipients with them at the target scene and would carry on mental conversations with them, pointing out various aspects of the sites. On some occasions, agents found their attention drawn to components of the scene they had initially overlooked, only to discover later that these features had been important parts of the percipient's description, almost as if the percipient's consciousness had guided their attention. And many participants indicated that they felt that the process was more akin to sharing a common experience than to "transmitting" information from one person to another.

To qualify for inclusion in the formal PEAR database, a trial was required to meet the following criteria:

1) The agent and percipient were specified to one another.

2) The date and time of the agent's target visitation were specified to the percipient.

3) The agent was present at the target within 15 minutes of the specified time and was consciously committed to his or her experimental role during that period.

4) Both agent and percipient produced verbal descriptions and completed appropriate descriptor response forms.

5) Both agent and percipient had adequate familiarity with the application and interpretation of the descriptor questions and with the general protocol.

6) Photographs, written descriptions, or other substantiating target information were available.

Trials that failed to meet these protocol criteria due a lack of adequate substantiating target information, evidence that one or both of the participants did not understand the application or interpretation of the descriptor questions, or the vulnerability of a trial to sensory cueing, were classified as "questionable" and were not included in the formal database although they were retained in the data pool. Another category of trials involved deliberate deviations from this formal protocol for exploratory purposes, such as not informing the percipient of the agent's identity, or not specifying the time of target visitation, and were identified in advance as "exploratory." All told, some 126 trials were categorized as questionable and 52 as exploratory.

As in many of the human/machine experiments, two methods of target selection were utilized: an "instructed" and a "volitional" mode. In the instructed version, used primarily for trials conducted in the Princeton area, the target location was selected randomly from a large pool of potential targets prepared previously by an individual not otherwise involved in the experiment, the directions to which were stored in separate randomly numbered sealed envelopes and maintained so that no agent, percipient, or experimenter had access to them. Prior to a given trial, the generation of a random number identified one of the envelopes, which then was delivered, still sealed, to the agent, who opened it after leaving the laboratory and then proceeded to the designated target location. In the volitional option, typically followed when the agent was traveling on an itinerary unknown to the percipient or the experimenters in a region for which no prepared pool existed, the agent simply selected a target from among the various local sites accessible at the time specified for the trial. Most PRP experiments were conducted as part of a "series" of from 3 to 20 independent trials involving the same pairs of participants and conducted over a relatively short period of time. The composite formal PEAR database of 653 trials comprised 88 such series, and was produced by a total of 39 agents and 59 percipients.

Another variation explored was the involvement of multiple percipients addressing a given target. In these experiments, the agents knew they were sharing their target experiences with more than one percipient and knew who they were, and the percipients were aware of the fact that there were others attempting to access the same targets. In these trials each percipient response to a given target was treated as an individual trial; as a result, the database contained several more perceptions than targets. A few other trials employed two people encoding the target scene as agents, but in these cases only one was the formal pre-designated agent and it was that individual's description that was used for scoring purposes; the second served solely for informal comparison.

These explorations extended over a period of some 25 years and involved five distinct phases, each of which addressed the development and investigation of a specific analytical judging procedure, or some variant thereof.[2] We shall review these efforts, summarize our findings, and speculate about their implications in the following few chapters.

References

[1] Russell Targ. *Limitless Mind: A Guide to Remote Viewing and Transformation of Consciousness.* Novato, CA: New World Library, 2004. p. 91.

[2] Brenda J. Dunne and Robert G. Jahn. "Information and uncertainty: 25 years of remote perception research." *Journal of Scientific Exploration, 17,* No. 2 (2003). pp. 207–241.

3

BINARY EPOCHS

We may not be able to get certainty, but we can get probability,
and half a loaf is better than no bread.
— C. S. Lewis[1]

As in the SRI and early Chicago experiments, the early PEAR PRP results were evaluated by independent human judges comparing each free-response perception with photographs of all of the targets in its particular series and assigning ranks to them, which were then subjected to statistical evaluation. A substantial database was acquired in this fashion, the results of which confirmed an overall anomalous acquisition of information at a level well beyond chance. Despite the impressive yield of these early experiments, concerns regarding the inefficiency of the ranking process and evident vagaries and possible subjective biases in the judges' interpretations, or even anomalous inputs on their part, predicated a more quantitative approach to data evaluation.[2] For example, the early procedures had been based on statistical evaluation of samples of only five to ten trials, where ranks beyond the first or second choices could have been assigned almost arbitrarily, but still have had a disproportionate impact on the outcomes. In addition, the ranking process had been starkly inefficient in that each trial had been represented by a single data point regardless of the amount of detail it might have contained.

To illustrate the issue, consider the following trial, where the target was the Westminster Cathedral, in London, England. The percipient was in Princeton, New Jersey, some 3500 miles away, and had produced the following description some four hours before the target was visited and photographed by the agent:

A rather vague, disconnected impression of a pattern sort of like nested arches. No other images at first, then a feeling of enclosure, or semi-enclosure, like a courtyard — but open on one side. Around the periphery a covered walkway with a solid wall on one side and open to the courtyard on the other. Pillars supporting roof. In center is some kind of object or group of objects in the center of some kind of setting involving foliage.

Westminster Cathedral: PRP target

Not at all clear or sharp. Also not clear if the whole scene is indoors or out. If indoors, brightly lit and an open, spacious feeling. Stone seems to be the dominant material or texture, but there's a sense of color as well. Also a feeling of oldness and somberness. Also feeling that I'm only picking up a portion of the scene — missing something important, maybe the thing in the center or maybe the whole image is part of a larger scene.

Clearly, there are some striking accuracies in this perception, such as the impressions of "nested arches," a "semi-enclosure, like a courtyard," "pillars supporting roof," "stone seems to be the dominant material or texture, but there's a sense of color as well," or "feeling of oldness and somberness." But there are also major components missing, such as the tall bell tower, and others that are minor or vague, such as a "group of objects in the center of some kind of setting involving foliage." It is as if the percipient had been gazing through a gauzy or cloudy filter, rather than through a well-focused lens, and even this had been restricted to a narrow angular scan of the target. Impressionistically, one might conclude that the above example provides a reasonably good portrayal of the target, but moving to a quantitative evaluation of the accuracy of such a description poses a substantial procedural challenge.

Since such characteristics seemed to pervade much of the data acquired under similar protocols, past and present, professional and personal, analytical and anecdotal, the issue of how to evaluate and interpret them in a more rigorous fashion became central to the course of our PRP program. The primary focus of the subsequent PEAR studies, therefore, was the development of an assortment of analytical judging procedures that were capable of rendering such free-response raw data into forms amenable to more precise quantification and thereby extracting the salient information embodied in each trial.

The development of these analytical techniques began with the replacement of the human judges with a standardized set of binary descriptor questions amenable to quantitative evaluation of the data and progressed through several phases and formats. Some 336 experimental trials were processed during the early stages of this work, categorized either as *ex post facto*, where existing experimental results were rendered into appropriate analytical encodings by laboratory staff, or as *ab initio*, where trials were encoded by the participants at the time of the trials. Two subsequent studies termed "FIDO" and "Distributive" that explored more complex encoding strategies will be addressed in the next chapter. Collectively, the overall 653-trial formal database was catalogued as follows:

1) *Ex post facto*–encoded database: 1976–1983 (59 trials, 7 series)

2) *Ab initio*–encoded database: 1980–1985 (277 trials, 42 series)

a) Early trials: 1980–1983 (168 trials, 29 series)
b) Later trials: 1983–1985 (109 trials, 13 series)
3) "FIDO"-encoded database: 1985–1989 (167 trials, 9 series)
4) "Distributive"-encoded database: 1992–1998 (150 trials, 30 series)

A. *Ex Post Facto*-Encoded Data

The potential statistical impact of inter-judge variability on the results of 27 early trials conducted in the Chicago area between 1976 and 1979[3,4] had been assessed by subjecting them to repeated re-judging by five separate individuals.[2] Approximately half of these trials showed strong consistencies between the ranks assigned by the original and subsequent judges, confirming the acquisition of significant extra-chance information. The others, however, received a diverse assortment of ranks, suggesting that some of the correct matches originally assigned to these trials had been somewhat arbitrary. This review also had confirmed the inherent inefficiency of an approach wherein the entire informational content of a given perception was reduced to a single data point, ordinal at best, in a small experimental series.

To alleviate some of these shortcomings, a standardized method was posed for quantifying the information content of the raw free-response data via a series of computer algorithms. To this end, a code, or alphabet, of thirty simple binary questions was established that could be addressed to all targets and all perceptions. The questions ranged broadly from factual to impressionistic:

1) Is any significant part of the perceived scene indoors?
2) Is the scene predominantly dark, *e.g.* poorly lighted indoors, nighttime outside, etc.?
3) Does any significant part of the scene involve the perception of height or depth, such as looking up at a tower, tall building, mountain, vaulted ceiling, or unusually tall trees, or looking down into a valley or down from any elevated position?

4) From the agent's perspective, is the scene well bounded, such as the interior of a room, a stadium, or a courtyard?

5) Is any significant part of the scene oppressively confined?

6) Is any significant part of the scene hectic, chaotic, congested, or cluttered?

7) Is the scene predominantly colorful, characterized by a profusion of color, or are there outstanding brightly colored objects prominent, such as flowers or stained-glass windows?

8) Are any signs, billboards, posters, or pictorial representations prominent in the scene?

9) Is there any significant movement or motion integral to the scene, such as a stream of moving vehicles, walking or running people, or blowing objects?

10) Is there any explicit and significant sound, such as an auto horn, voices, bird calls, or surf noises?

11) Are any people or figures of people significant in the scene, other than the agent or those implicit in buildings or vehicles?

12) Are any animals, birds, fish, or major insects, or figures of these, significant in the scene?

13) Does a single major object or structure dominate the scene?

14) Is the central focus of the scene predominantly natural, that is, not man-made?

15) Is the immediately surrounding environment of the scene predominantly natural, that is, not man-made?

16) Are any monuments, sculptures, or major ornaments prominent in the scene?

17) Are explicit geometric shapes — for example, triangles, circles, or portions of circles (such as arches), or spheres or portions of spheres (but excluding normal rectangular buildings, doors, windows, and so forth) — significant in the scene?

18) Are there any posts, poles, or similar thin objects, such as columns, lampposts, or smokestacks (excluding trees)?

19) Are doors, gates, or entrances significant in the scene (excluding vehicles)?

20) Are windows or glass significant in the scene (excluding vehicles)?

21) Are any fences, gates, railings, dividers, or scaffolding prominent in the scene?

22) Are steps or stairs prominent (excluding curbs)?

23) Is there regular repetition of some object or shape, for example, a lot full of cars, marina with boats, or row of arches?

24) Are there any planes, boats, or trains, or figures thereof, apparent in the scene, moving or stationary?

25) Is there any other major equipment in the scene, such as tractors, carts, or gasoline pumps?

26) Are there any autos, buses, trucks, bikes, or motorcycles, or figures thereof, prominent in the scene (excluding the agent's car)?

27) Does grass, moss, or similar ground cover compose a significant portion of the surface?

28) Does any central part of the scene contain a road, street, path, bridge, tunnel, railroad tracks, or hallway?

29) Is water a significant part of the scene?

30) Are trees, bushes, or major potted plants apparent in the scene?

The percipients' and agents' binary responses were entered into a database manager as strings of 30 bits (1 for yes, 0 for no) and submitted to an assortment of analytical scoring algorithms to compare the perception features with those of the actual targets. The statistical merits of the perceptions were then evaluated by an assortment of computerized analytical ranking procedures that scored each transcript against all of the targets in the pool and then ranked them in order of descending score.[5] While still requiring a ranking procedure, this descriptor-based process had the advantages that such rankings could proceed from a more standardized analytical approach and that many more alternative targets could be ranked by the computer than by a human judge.

As a first test of this approach, one series of eight trials from the Chicago database was encoded *ex post facto* into the binary format by five independent encoders. Reassuringly, most of these responses were found to be in close agreement with each other, and the computer-assigned ranks of the better trials were consistent with those of the original human judges while those of the weaker trials were comparably equivocal.

With the scoring methods thus qualified, 35 new trials were performed following the same protocol used in the earlier experiments, but now the targets and perceptions were descriptor-encoded *ab initio* by the agents at the target sites and by the percipients after completing their free-response descriptions. Although the results of these new trials were not as striking as those of the *ex post facto*–encoded data, they were still highly significant statistically. More encouragingly, the general agreement among the various scoring algorithms confirmed that this analytical methodology was indeed capable of providing reliable quantification of the intrinsically impressionistic remote perception data.

To obviate the possibility that the particular list of descriptors employed somehow could process even random inputs to apparently significant scores, a control exercise was undertaken wherein artificial "target" and "perception" data matrices of the same size as the experimental ones were constructed from the output of a random event generator. The same computational schemes were applied to various combinations of these, both with each other and with the true data, with results that were all well within chance expectation.

The remaining early human-judged trials were then also transcribed *ex post facto* into the new descriptor format. Five individuals who were blind to the correct target/perception matches performed the encoding, and the response to each descriptor question was determined by majority vote. The total *ex post facto*–encoded data subset comprised a total of 59 trials, and included all the original human-judged trials from Chicago and PEAR that met formal protocol criteria and had adequate target documentation to permit such retrospective encoding. Reassuringly, the results of these re-encoded trials also were highly consistent with those that had been obtained in the earlier human judging process.

B. *Ab Initio*-Encoded Database

Beyond its evident success in ranking the trials in any given experimental series, the descriptor-based scoring method offered another desirable and powerful capability, *i.e.* the direct calculation of the statistical merit of individual trial scores or groups of scores. By scoring every perception in the database against every possible target except the correct ones, it was possible to construct an empirical chance distribution of mismatched scores, the distribution of which displayed classical Gaussian features that could serve as a statistical reference for assessing the extra-chance likelihood of the matched perception/target scores.[6]

By 1983, the original 59 *ex post facto*–encoded trials had been supplemented by an additional 168 formal trials that were encoded *ab initio* by the participants at the time of the trials. During this period, several variations of the scoring method also were elaborated, each of which consisted of calculating a score for each trial based on the proportion of matches and mismatches in the percipient and agent responses to the thirty descriptor queries. The simplest of these variants merely counted the number of descriptors answered correctly and divided that by the total number of descriptors. Others employed various weighting factors that took into account the *a priori* probabilities for each descriptor response. For example, since more targets tended to be outdoors than indoors, a correct positive response to the query "Is the scene indoors?" was assigned a greater weight than a correct negative response, and its incremental contribution to the total score was proportionately larger. Various normalization procedures also were explored, such as dividing the absolute score for a given trial by the maximum score that would have been achieved had all thirty target and perception descriptor responses agreed.[7]

In addition to the standard "yes" and "no" options, the descriptor-response check sheets also contained a column labeled "unsure" that permitted participants to indicate any ambiguities they might experience in relating their subjective impressions in strictly binary terms. These "unsure" responses were not included in the binary calculations, but they provided the basis for investigating the potential benefits of several ternary-based algorithms that explicitly included this option.[8] Seven such ternary recipes were explored, all of which showed good internal consistency, but none of which indicated any substantial advantage over the binary calculations. Given their added computational complexity, they were abandoned, and subsequent study was limited to five of the original nine binary-based methods.

Comparisons of the distributions of properly matched scores with empirical chance distributions consisting of all the mismatched scores, computed by the same recipe, permitted calculation of simple z-scores that indicated the statistical likelihood of the correctly matched scores containing significantly more informational correspondences with the targets than would be expected by chance. By this criterion, the 168 *ab initio*–encoded trials proved to be highly significant, with z-scores in the range of 4.6 ($p = 2 \times 10^{-6}$), depending on the algorithm deployed. Although this was somewhat less impressive than the z-score of 5.8 ($p = 3 \times 10^{-9}$) obtained for the 59 *ex post facto*–encoded trials, it was assumed that this reduced effect could be attributed at least in part to the fact that the development of the descriptor questions and analytical techniques had been based on the content of the earlier trials.

The cumulative deviation graphical format that had proven so incisive and comprehensive in illustrating the evolution of our human/machine experimental results also served to summarize the accumulating PRP data effectively, albeit with some analytical conversions.[9] Figure III-1 displays such representation for the database of all 336 formal trials.

One of the most reassuring features of these results was the consistency of anomalous yield across the five diverse binary scoring schemes pursued. Regardless of the algorithm employed, the composite results indicated highly significant increments

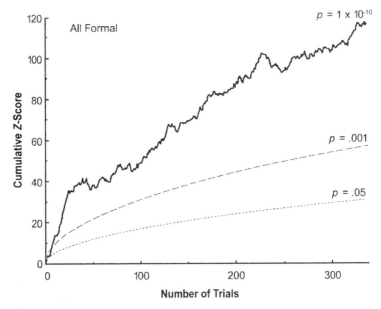

Fig. III-1. Cumulative *z*-score graph of all formal binary PRP trials (*cf.* Ref. 9).

of anomalous information in the matched scores that were not present in the mismatched score distributions constructed from the same raw data. Even the null results of the exploratory trials were informative in their indication that certain features omitted from the standard protocol in these excursions, *i.e.* the percipients' knowledge of the agent or of the time of target visitation, might be important factors in the anomalous acquisition of target information.

Despite its demonstrated advantages, however, the binary analytical judging processes introduced certain imperfections of their own. For example, the forced "yes" or "no" responses were limited in their ability to capture the overall ambience or context of a scene, or the nuances of subjective or symbolic information that might be detected by human judges. Furthermore, while restricting the extracted information to the 30 specified binary descriptors simplified the reporting task for the participants and the analytical task for the experimenters, it precluded utilization of other potentially relevant features in the transcripts, including such subjective impressions as specific colors, textures, architectures, or other unusual details not covered by the questions, so that percipients occasionally reported that they felt overly constrained by the questions.

These shortcomings were partially offset by the continued requirement that percipients first generate free-response descriptions to which the descriptor responses were then applied, a procedure intended to retain the spontaneity of the PRP experience as well as to preserve the raw data in a suitable format for further study. Despite the reduction in the yield from that of the earlier trials, the *ab initio*–encoded data were still highly significant, and the sacrifice of some of the subjective impressions of the earlier trials was deemed a reasonable price to pay for the advantage of more incisive quantitative measurement. Nonetheless, it gradually became evident that many participants were tending to limit their attention and descriptions to those features that they now knew were specific to the anticipated questions.

These limitations notwithstanding, the overall advantages of the analytical judging techniques encouraged their further development, beginning with a comprehensive evaluation of the effectiveness of the individual descriptors in constructing the trial scores. From this it was determined that the entire group of descriptors, originally selected by some combination of anecdotal experience and intuition, actually comprised a reasonably uniform set in terms of their effectiveness in quantifying informational bits across a broad range of target types. None was found to be extremely effective; none was seriously deficient. Subdivision of the descriptors into classifications of natural vs. man-made, objective vs. impressionistic, permanent vs. transient, and indoor vs. outdoor, also revealed no significant differences in effectiveness. The interdependence among the various descriptors, *e.g.* that outdoor scenes were less likely to be confined, or that indoor scenes seldom involved airplanes or road vehicles, was also explored by a variety of statistical methods, all of which confirmed that whereas such correlations might blunt the incisiveness of the full descriptor net somewhat, they did not compromise the validity of the overall results.[10] With these reassurances in hand, an additional 109 new *ab initio* trials were conducted. Bemusingly, the statistical yield of this sub-set (referred to as the later *ex post facto* database) proved to be indistinguishable from chance!

C. Secondary Correlations

Over the course of these analytical excursions, the importance of a number of secondary parameters was also explored, the results of which are summarized in Table III-1. Beyond the instructed vs. volitional determination of target designation described earlier, these included the effectiveness of employing multiple percipients; seasonal influences; geographical areas; and distance and time dependences. Since the protocols, descriptor questions, and scoring algorithms in the *ex post facto* and two *ab initio* subsets were virtually identical, they could legitimately be combined to provide a larger database for the other structural segmentations listed in the Table.

TABLE III-1

Summaries of Binary PRP Data

Subset	# Trials	Effect Size*	z-score	Prob. (1-tailed)	# Trials $p < .05$*	% Trials $p < .05$	% Trials $p < .50$
All Formal Trials	336	.347	6.355	1×10^{-10}	44 (8)	13 (2)	62
Ex Post Facto	59	.754	5.792	3×10^{-9}	14 (2)	24 (3)	75
Ab Initio (all)	277	.263	4.378	6×10^{-6}	31 (5)	11 (2)	59
Ab Initio (early)	168	.354	4.582	2×10^{-6}	24 (4)	14	61
Ab Initio (later)	109	.124	1.291	.098	7 (2)	6 (2)	60
Instructed	125	.516	5.771	4×10^{-9}	23 (5)	18 (4)	65
Volitional	211	.244	3.549	2×10^{-4}	25 (3)	12 (1)	60
Single Percipient	216	.382	5.613	1×10^{-8}	34 (6)	16 (3)	60
Multiple Percipient	120	.312	3.416	3×10^{-4}	12 (3)	10 (3)	63
Summer Trials	244	.363	5.663	7×10^{-9}	35 (5)	14 (2)	65
Winter Trials	92	.315	3.017	1×10^{-3}	13 (2)	14 (2)	57
Non-Formal Trials	75	−.046	−0.399	.655	3 (4)	4 (5)	44

*Effect sizes were calculated by dividing the z-scores for each database by the square root of the number of trials in that subset, and thus indicate the average z-score per trial.
( ) indicates results opposite to intention.

Of the 336 formal trials in the binary database, 125 followed the instructed protocol where the target was selected at random from a pre-existing pool, and 211 utilized the volitional protocol wherein the agent selected the target by some *ad hoc* criterion. (Another 54 trials were classified as exploratory and 21 as questionable.) Note that the overall effect size and the composite z-score for all the instructed trials (.516 and 5.771, respectively) were substantially larger than for those of the volitional trials (.244 and 3.549, respectively), but it is important to keep in mind that most of the high-scoring *ex post facto* data were from instructed trials. (In contrast, the subsets of 94 instructed and 183 volitional trials that comprised the *ab initio* database were statistically indistinguishable, with z-scores of 3.12 and 3.15, respectively.) There were, however, nearly twice as many *ab initio* volitional trials as instructed, so that the average effect sizes of these two groups (.322 for the instructed trials, and .233 for the volitional) actually indicated an enhanced effect in the instructed trials, even with the *ex post facto* data excluded. A substantial percentage of the formal trials (13%) exceeded the $p < .05$ chance probability levels, while there was a smaller number of scores in the negative tail of the distribution than expected by chance (2%).

The differences between the instructed and volitional trials is informative since the less formal nature of the target selection process in the volitional trials, or the possibility of the agent's knowledge of the percipient's personal preferences or target response patterns, might be suspected to have influenced the target selection and representation so as to introduce an undue bias in favor of the volitional trial scores. Although there was indeed a significant difference between the results of these two subsets, it was actually the *instructed* subset that repeatedly produced the larger effect sizes, so that any concern that the target selection process employed in the volitional trials might have contributed to artificial enhancement of the results appeared to be unfounded.

Examination of the single vs. multiple percipient subsets indicated no advantage to having more than one percipient address a given target, at least within the protocols employed. Of the 336 formal trials, 216 were generated under the standard protocol wherein a single percipient attempted to describe the location of a single agent. In the remaining 120 trials, two or more percipients addressed the same target. The number of percipients addressing a given target ranged from two to seven, and each perception was scored as a separate trial against its appropriate target. In all the multiple-percipient trials the agents knew who the various percipients were and the percipients were aware that others were involved in the experiment, although they did not always know their identities. The various percipients always were separated spatially from each other and, in most cases, attempted their perception efforts at different times. These two subsets showed many strong similarities, with only a slightly higher yield for the single-percipient trials. Comparisons of datasets grouped by time of year or by geographical location (Princeton targets vs. targets elsewhere) also showed nearly identical effect sizes.

D. Binary Summary

By the close of this phase of the program in 1985, a number of informative general conclusions had emerged:

- There was general agreement between the results of the various analytical methods and those of the impressionistic assessments by human judges, particularly for the perceptions of higher statistical merit.
- Although the various scoring recipes produced somewhat different scores for individual trials, the composite statistical yield was uniformly highly significant and relatively insensitive to the particular method employed.
- The use of ternary descriptor responses, wherein participants were offered the option of "passing" on a given descriptor, did not yield sufficiently more consistent or accurate results compared to the binary methods to justify the added computational complexity.

- Defining a "universal" target pool containing a sufficiently large number of actual targets made it possible to calculate generalized *a priori* descriptor probabilities that could be used for scoring any individual perception efforts in the database, regardless of its particular local series pool.
- Calculation of the statistical merit of individual perception efforts by reference to an empirical chance distribution derived from a large number of deliberately mismatched targets and perceptions, proved to be a more powerful strategy than the computerized analytical ranking within individual small series.
- The 30 descriptors, originally chosen through a combination of empiricism and intuition, while clearly non-independent, nonetheless displayed a reasonably flat profile of effectiveness in building the scores of the significant transcripts.
- The trials generated under the instructed protocol produced larger effect sizes and statistical yields than those following the volitional protocol.
- Trials grouped according to time of year or geographical region showed virtually identical effect sizes.
- The 75 informal trials produced essentially chance results.

One further analysis of these data was prompted by a critical challenge that since the agents and percipients knew one another, their descriptor responses could be influenced by shared response biases that could artificially inflate their trial scores and compromise the analytical results. As mentioned, all of the scoring methods calculated trial scores that were based on the local *a priori* descriptor probabilities associated with each particular data subset. When those local probabilities were used to score a given subset using a given scoring method, the empirical chance distributions resulting for different subsets appeared to be statistically indistinguishable and it would seem to follow that a single empirical chance distribution, namely the one deriving from the largest commensurate assembly of formal data, could be used as the reference standard for any subset, provided that the subset's trial scores were computed using its own local *a priori* probabilities. Unfortunately, this uniformity of chance distributions was only approximately correct.

A potential mechanism whereby local variations in the *a priori* probabilities among the different subsets of the database could potentially produce artificially inflated, or deflated, scores in the matched-trial distributions relative to the off-diagonal population of mismatches was also considered.[7] For example, a given percipient/agent pair might conceivably happen to share a similar encoding style, such as a tendency to respond affirmatively to ambiguous features, or to share particular preferences for certain descriptors, which could result in their trials having responses that were more closely correlated than those of the mismatched scores constituting the reference distribution. Similar biases might also have arisen from geographical or seasonal variations, among other possible causes.

Since the apparent indistinguishability of the chance distributions for a number of large data subsets could not be guaranteed theoretically, it was necessary to verify empirically that the overall results were not in fact spuriously inflated by such biasing mechanisms. The possible influence of idiosyncratic individual patterns of response in agent and percipient encoding styles was examined using the data produced by the 29 agent/percipient pairs who had contributed five or more trials to the composite binary database. Collectively, these pairs were responsible for 274 of the 336 formal trials. The results of this test for possible local biasing are shown in Figure III-2, which displays an array of cumulative deviation graphs over these 274 trials, where the individual plotted points are the *z*-scores accumulated by each of the 29 participant pairs based on three different calculation methods.

The "non-local" method is simply the composite *z*-score achieved in this subset of trials by given percipient/agent pairs. The "local alpha" score is derived by scoring each percipient/agent pair contribution on the basis of its own internal *a priori*

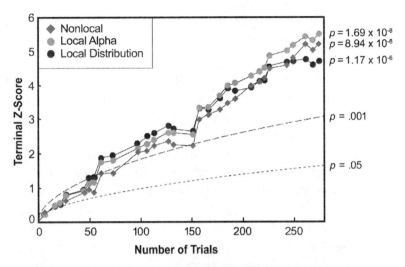

Fig. III-2. Terminal *z*-scores achieved by 29 percipient/agent pairs for three alternative PRP scoring techniques (*cf.* Ref. 9).

probabilities, but still referring these scores to the overall chance distribution. The "local distribution" calculation removes all reference to global distributions, and along with it any possibility of a local-biasing effect, by scoring each agent/percipient pair trial using its own local *a priori* probabilities, and comparing against its own local mismatch distribution.

This comparison made it evident that these methods are not statistically distinguishable and that any inflation or deflation of the overall effect due to local biasing was less than the inherent statistical uncertainty of the scoring procedures. Thus, we could comfortably conclude that encoding artifact was not a significant contributor to the experimental results.

The rank-ordered effect sizes obtained by each of the 28 percipients and 15 agents who contributed more than one trial to the binary database were also examined. Some 25% of the percipients, 40% of the agents, and 21% of the percipient/agent pairs produced statistically significant overall results, whereas only 5% of each group would be expected to do so by chance. All but two percipients and two agents generated net positive effects, compared to the 50% chance expectation, and of these four individuals, three produced positive results when functioning in the alternate roles.

In one other reassuring calculation, a separate data subset, consisting of only the first trials from each of the 38 individual percipients who contributed to the formal database, was examined to explore the possibility that the composite yield might have been distorted by large databases produced by any given percipient. Despite the small size of this group of trials, the results displayed the same statistical consistency as the full database (*cf.* Figure III-1), achieving a highly significant composite *z*-score of 3.89 (Fig. III-3), confirming that the success of the overall results was not attributable to exceptional performance by a few participants, and demonstrating once again the series position pattern we had observed in our human/machine experiments.

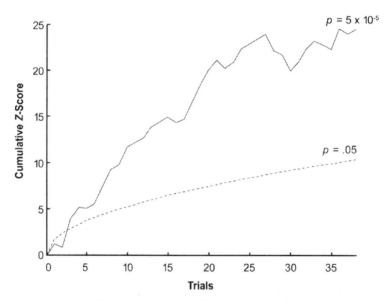

Fig. III-3. PRP cumulative *z*-score, 38 percipients, first trials only (*cf.* Ref. 9).

Notwithstanding these reassurances of the reliability of the analytical technique, the severely diminished effect size in the later *ab initio*–encoded trials remained problematic, and prompted a new phase of investigation aimed at better understanding the cause of this deterioration, and at attempting to recover the stronger yields obtained in the original experiments.

References

[1] Clive Staples Lewis. *Christian Reflections.* Walter Hooper, Ed. London: Bles, 1967. p. 111.

[2] Brenda J. Dunne. "Precognitive Remote Perception: A Critical Overview of the Experimental Program." Master's Thesis, Committee on Human Development, University of Chicago, Chicago, IL. June, 1979.

[3] Brenda J. Dunne and John P. Bisaha. "Precognitive remote viewing in the Chicago area: A replication of the Stanford experiment." *Journal of Parapsychology, 43* (1979). pp. 17–30.

[4] John P. Bisaha and Brenda J. Dunne. "Multiple subject and long-distance precognitive remote viewing of geographical locations." In Charles T. Tart, Harold E. Puthoff, and Russell Targ, eds. *Mind at Large: IEEE Symposia on the Nature of Extrasensory Perception.* New York: Praeger Special Studies, 1979. pp. 109–124. (Reprinted: Charlottesville, VA: Hampton Roads Publishing Company, 2002. pp. 98–111.)

[5] Robert G. Jahn, Brenda J. Dunne, and Eric G. Jahn. "Analytical judging procedure for remote perception experiments." *Journal of Parapsychology, 44* (1980). pp. 207–231.

[6] Brenda J. Dunne, Robert G. Jahn, and Roger D. Nelson. "Precognitive Remote Perception." Technical note PEAR 83003. Princeton Engineering Anomalies Research, Princeton University, School of Engineering/ Applied Science, Princeton, NJ. August 1983.

[7] Brenda J. Dunne, York H. Dobyns, and Susan M. Intner. "Precognitive Remote Perception — III: Complete Binary Data Base with Analytical Refinements." Technical note PEAR 89002. Princeton Engineering Anomalies Research, Princeton University, School of Engineering/Applied Science Princeton, NJ. August 1989.

[8] Robert G. Jahn, Brenda J. Dunne, Roger D. Nelson, Eric G. Jahn, Todd Aaron Curtis, and Ian A. Cook. "Analytical Judging Procedure for Remote Perception Experiments — II: Ternary Coding and Generalized Descriptors." Technical note PEAR 82002. Princeton Engineering Anomalies Research, Princeton University, School of Engineering/Applied Science. Princeton, NJ. 1982.

[9] Brenda J. Dunne and Robert G. Jahn. "Information and uncertainty: 25 years of remote perception research." *Journal of Scientific Exploration,* *17,* No. 2 (2003). pp. 207–241.

[10] Jensine Andresen. "Statistical Tests of Experimental Remote Perception Trials." Undergraduate thesis, Department of Civil Engineering, School of Engineering/Applied Science, Princeton University, Princeton, NJ. 1984.

4

A BRIDGE TOO FAR

If we knew what it was we were doing,
it would not be called research, would it?
— Albert Einstein[1]

Although the binary analytical judging techniques had shown that several versions of them could capture the essence of PRP information in a manner amenable to standardized quantification, the essentially null results of the later *ab initio* trials were troublesome and demanded further investigation. The frequent participant complaints that the forced binary responses seemed somewhat inhibitory and could only incompletely represent their target impressions suggested that this format might have been eroding the results. It was indeed evident that many of the target scenes contained ambiguous features that could not be answered adequately with simple "yes" or "no" responses. For example, an agent might be indoors but looking out a window at an outdoor scene, and thus be unsure whether to characterize the scene as indoors or outdoors. Or a feature might have captured the agent's attention during the target visitation but not have been an integral component of the scene itself, such as a brief conversation with a passerby in an otherwise unpopulated area, complicating the response to the question "Are people present?" A similar problem arose in percipients' efforts to extract details from a perception that emerged as a less than coherent stream of consciousness, somewhat akin to fragments of dream imagery. Since our initial explorations of ternary descriptors that permitted participants to "pass" on those questions that seemed ambiguous or irrelevant had failed to improve the results, we decided to pursue a more elaborate approach.

"Are people present?"

A. FIDO

To this end, a quaternary descriptor response version was devised, playfully termed FIDO, an acronym for "Feature Importance
Discrimination Option." This new format offered participants four
response options for each descriptor: a rating of 4 to identify a
feature that was a clearly dominant component of the scene; 3 to
denote a feature that was present, but not particularly important
or central; 2 to indicate uncertainty as to the presence or absence
of the feature; and 1 to record the definite absence of the feature.
Since implementation of the FIDO method required rewording of
the descriptors, combination of the FIDO trials with the earlier
databases was not feasible, but the rewording exercise did provide an opportunity to clarify or redefine some questions that had
posed occasional interpretational difficulties.

A slightly revised set of 32 questions was developed and calibrated by having several people visit a variety of test scenes and
encode them with the new quaternary descriptors, after which
their responses were compared for consistency. Reassured that
these new descriptors provided reasonably coherent responses,

we undertook a fresh body of experiments. In all other respects, the protocol remained the same, save that data were now generated on an individual trial-by-trial basis, rather than in series of varying lengths. This FIDO program again utilized five alternative scoring algorithms, and it ran for four years, providing a total of 167 trials.[2]

To our deep disappointment and bemusement, the sought enhancement of the trial scores was not achieved. Although the results of these different scoring methods all displayed close concurrence, their statistical yield by the matrix/mismatching technique continued to be reduced even further from that of the earlier data to a marginal composite z-score of only 1.74 (p = .04), and a corresponding average effect size only about a fifth that of the *ex post facto* subset. And while a few individual FIDO trials, such as those illustrated in the next chapter, showed clear percipient/agent correlations, and more than 50% of the FIDO trials achieved scores above chance in all of the alternative calculations, neither the overall results nor the numbers of significant trials substantially exceeded chance expectations.

Thus, this FIDO protocol clearly had not achieved its goal of enhancing the overall PRP yield, nor had it improved the potential sensitivity to subtle or ambiguous nuances in the data. To the contrary, the internal consistency of this collection of results suggested that the decreased yield was not directly due to inadequacies in the FIDO scoring algorithms, *per se*, but to a more generic suppression of the anomalous information channel somehow imposed on the experimental protocol by the FIDO format. This suspicion was reinforced by an exercise in which an independent human judge was asked to rank the agreements of the agents' free-response transcripts with their FIDO-coded response sheets. While this ranking effort was admittedly subjective and arbitrary, and was complicated by the varied lengths of transcripts and the presence or absence of drawings, photos, or other illustrative material, this judge determined that 162 of the 167 targets showed reasonably good correspondences between the agents' verbal descriptions and their descriptor-coded responses. A similar exercise was performed on the percipients' encodings of their transcripts, with comparable results. The FIDO descriptors themselves thus seemed adequate for capturing both the target information and the percipients' imagery, forcing us to conclude that the source of the diminishing returns lay elsewhere.

B. Distributive Scoring

Hints of possible co-operator and gender-related trends like those found in our human/machine studies (*cf.* Chapter II-13) also had been noted in the PRP data, but the existing pool of contributing percipients and agents had been too small and disproportionately balanced to determine whether such gender pairing might be responsible for the diminished yield. Another new remote perception protocol therefore was developed to explore this possible correlation more systematically, using a balanced pool of same- and opposite-sex participant pairs, each contributing an equal number of trials.

This protocol required each percipient/agent pair to generate one series consisting of five trials, and then another five-trial series with their roles reversed. A concern expressed by a critic, that providing feedback to participants at this conclusion of each trial could introduce a possible bias in subsequent trials, was addressed by withholding feedback until all five trials of a series were completed, and each target selected from the pool in the instructed experiments was replaced before the next trial. Finally, to preclude any possibility of shared response biases, all analyses were based solely on local subset comparisons within a given series.

Despite the greater flexibility in responding to the descriptors offered by FIDO, some participants were still commenting that these constrained their ability to depict their experiences adequately. At this point, we devised yet another set of descriptors that permitted participants to respond to each of 30 features on a distributive scale of 0 to 9 to indicate the relative prominence of each feature yet more precisely. The individual trial scores were calculated from 10×10 matrices that cross-indexed and assigned values to every possible pair of percipient/agent 0–9 descriptor rankings. Six different scoring recipes were explored, for each of which the sum of the individual descriptor scores constituted the total score for a given trial. As in the prior methods, overall statistical results for each series were evaluated by constructing 5 × 5 matrices comprising the scores of every target matched against every perception, and the scores of the five matched trials were compared with those of the 20 mismatched scores to determine the statistical merit of each series.

Twelve participant pairs, eight of whom produced at least two series together with their percipient/agent roles reversed, generated thirty experimental series using this distributive protocol, comprising a total of 150 trials. Once again, we found reasonably good agreement among the six scoring recipes, but the overall results of these trials were now completely indistinguishable from chance. Nonetheless, there were still a few perceptions, such as the two included below, that produced correspondences with their targets that were unlikely to occur by simple guessing, although each of them displayed the usual mixture of accuracy and cognitive overlay common to so many remote perception efforts.

In the first example, which illustrates the very brief written descriptions typically obtained in these distributive trials, the target was a supermarket in Princeton, described by the agent as:

Walking around supermarket, pushing a cart, shopping for lab groceries — coffee, etc. Busy, colorful, noisy. Selected large bunch of sunflowers. (No photograph or drawing accompanied the target description.)

The percipient, approximately 20 miles away, described the scene 1½ hours before its designation:

Rows or aisles, very organized as in a shopping center. Tile or linoleum floor. Music playing. Many small items such as cans of food or packages of some sort.

In another, retrocognitive, example, the target was a side section in the main Princeton University Chapel. Here, the agent's description was much more lengthy and detailed, and included a sketch along with a considerable amount of subjective reaction to the scene:

Wow. This place is creepy. Warm, yes, but dark. Like something from a fairytale. One spotlight on the mini-altar. To the left, the crucifix. Ceiling is wooden. All walls are stone. Marble? Feels like a dungeon. A tour group from behind sounded like ghosts. The only link to reality are the bird calls outside, beyond the wooden doors. Center-gold-encased bible/holy communion. Right — red glass tube for holding a candle. The base is gold, and there is no candle inside. Below, from a kneeling position, there is a picture of Mary in a turquoise + gold veil, tearless crying .... Super elaborate stained glass outside. Cannot see from room. Halogen

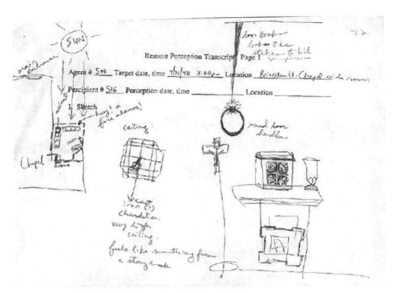

Chapel: PRP target (agent sketch)

bulbs??? (from chandelier) Also, a ring, hanging from the bottom, also looks like cast iron. Tour group left. All would be quiet, except for those birds outside. Very loud. Light from sun through a crack in the door looks like spectrum pattern on the floor. Door looks like stakes to kill vampires.

The percipient was in Atlanta, Georgia, and recorded the following perception eight hours *after* the target visit:

A church. Inside the benches and floor are dark wood. It's small, only about 20-30 rows of benches. (Agent) is about 2/3 of the way back, looking ahead at the altar and the large stained glass window shaped like [sketch]. The stained glass doesn't let much light in, the church has a general dimness in it. There are only about 10 other people in the church — sitting, praying, and wandering about.

These provocative examples notwithstanding, the low statistical resolution in defining the local empirical chance backgrounds (a consequence of the small size of the scoring matrices) made calculation of individual trial z-scores virtually meaningless. And given the lack of any overall statistical yield in these data, it was not possible to ascertain whether there was any evidence of co-operator or gender effects — the question that had originally prompted this exploration. In short, with the exception of a few encouraging trials, the remote perception phenomenon itself seemed once again to have been suppressed.

Contemplating this paradox, we considered a number of subtler, less quantifiable factors that might have had an inhibitory effect on the FIDO and distributive experiments, some having to do with the laboratory ambience in which they had been conducted. For example, during the period in which the FIDO data were being generated, the staff had been distracted by preparation of a systematic refutation of an article critical of PEAR's earlier PRP program.[3,4] Although most of the issues that had been raised in that article were irrelevant, incorrect, or already had been shown to be inadequate to account for the observed positive effects in the earlier data, this challenge deflected attention from, and dampened the enthusiasm for, the experiments in progress at that time. Beyond this, in order to forestall further such specious challenges, it had led to the imposition of additional pettifogging constraints in the design of the distributive protocol. Although it is difficult to quantify the influence of such issues, in the study of consciousness-related anomalies where unknown psychological factors appear to be at the heart of the phenomena, they cannot be dismissed casually.

In what might be a more productive line of speculation, we found ourselves wondering why, in spite of all our efforts to relax the constraints on the descriptor queries, our participants continued to find them inhibiting. Was it possible that our emphasis on quantifying the importance of a given descriptor, rather than merely acknowledging its presence or absence, was itself an inhibiting factor?

References

[1] Albert Einstein. *Out of My Later Years.* New York: Citadel Press, 1941.

[2] Brenda J. Dunne and Robert G. Jahn. "Information and uncertainty: 25 years of remote perception research." *Journal of Scientific Exploration, 17,* No. 2 (2003). pp. 207–241.

[3] George P. Hansen, Jessica M. Utts, and Betty Markwick. "Critique of the PEAR remote-viewing experiments." *Journal of Parapsychology, 56,* No. 2 (1992). pp. 97–113.

[4] York H. Dobyns, Brenda J. Dunne, Robert G. Jahn, and Roger D. Nelson. "Response to Hansen, Utts, and Markwick: Statistical and methodological problems of the PEAR remote viewing [sic] experiments." *Journal of Parapsychology, 56,* No. 2 (1992). pp. 115–146.

5

NON-LOCAL MIND

*... the localization of the personality, of the conscious mind,
inside the body is only symbolic, just an aid for practical use.*
— Erwin Schrödinger[1]

Whatever may have caused the lack of success of the FIDO and
Distributive exercises in enhancing the PRP yield, the data made
it evident that it was *not* the spatial or temporal separations between the targets and the percipients. As in the earlier analyses,
no correlations could be found with these physical parameters.
The possibility that an individual can obtain information about a
remote location in a manner that does not depend on the usual
senses is difficult enough to reconcile within the framework of
current scientific representation. That a person might do so *before* the site has been visited, or even specified, seems still less
credible. Yet, the following two transcripts drawn from the FIDO
database, despite the usual sprinkling of inaccuracies, omissions,
and misimpressions, refute any diminution of the remote perception process due to large spatial and temporal separations.

In the first case, the agent was in Cornwall, UK, viewing
Launceston Castle. He did not provide a verbal description of the
target, but submitted the two photographs shown on the following page. The percipient was in Pompano Beach, Florida, over
4000 miles away, and described the target approximately 31 hours
before the agent visited it:

*Outdoor, urban imagery — buildings, streets with traf-
fic, noise, people moving about. Cloudy, overcast, possibly even drizzling. Agent standing with back to large
gray building looking at a circular structure or building*

Launceston Castle, Cornwall: PRP target

of unusual design, possibly with some kind of spi-
ral motif twining around it. (Sort of like Guggenheim
museum in NY) No windows noticeable, but an un-
usual large "entryway" surrounded by ornate decora-
tion of some kind. There's a regularly spaced row of
stone posts of some kind — can't place it relative to
the main structure. The main structure is of gener-
ally circular shape, somewhat higher than it is wide,
made of light colored material, probably stone, but
not plain concrete — more like a decorative stone like
marble or something like that. Possibly a pale "pastel"
color, pinkish. The whole thing could be a large sculp-
ture of some kind — scale not clear.

In the second example, the agent was at a sidewalk café in
Balaton, Hungary, and the percipient was in Burlington, Vermont,
a distance of approximately 5000 miles, and the perception was
generated nine days prior to the agent's visit. The agent did not
provide a photograph, but described the target verbally, as follows:

I went to a sidewalk café with the students at around
1530 hours. We drank beer and wine and sat outside
under trees. There were a lot of German tourists

234

around us. I then went to their summer house and drank more beer and wine.

The percipient's transcript read:

I see (the agent) sitting at a table, in an outdoor café or at a brightly lit indoor café table. He is with 2 or three others, drinking something. (tea or beer?), and talking; perhaps about this PRP process. — There are leaves and vegetation around, perhaps they're sitting among trees — or there are lots of plants around. The spirit is lively and the people are having a good time. The ground surface is inlaid stone of some kind, perhaps cobblestone.

When agglomerated with many other striking examples of accurate long-distance, off-time perceptions that were evaluated by our various other analytical scoring methods, the body of evidence for the lack of dependence of the remote perception phenomena on the physical time and distance intervals may be the most noteworthy aspects of the results, and the most seminal in illuminating the underlying characteristics of the anomalous communication process. The empirical demonstration that such capabilities are accessible to ordinary people under controlled conditions further challenges any attempts to apply conventional physical models. For example, if some kind of electromagnetic or geophysical wave propagation were responsible for these effects, one would expect to see indications of an inverse dependence of the fidelity of the signal on the intervening distance; the insensitivity to intervening time is even more implacable in any classical physical model.

The spatial distances in the full body of our PRP experiments ranged from less than one mile to several thousand miles, and the temporal separations from several days before to several days after target visitation. When all of the binary PRP scores were plotted as functions of the distance or time displacements, simple regression calculations confirmed that there were no significant deteriorations of the effects with increased distance or time. As an alternative criterion, when the data were segregated into subsets of the more extreme spatially and temporally displaced trials and those more proximate, the average effect sizes of the former remained statistically indistinguishable from those of the latter.[2] Figures III-4 and III-5 display the regression fits to the 336-trial formal binary data subset.

As mentioned in the previous Section, these distance- and time-independent PRP results prompted a testable hypothesis that our human/machine anomalies might display similar non-local characteristics. There too, significant intention-correlated mean shifts were observed that were statistically indistinguishable from those in the local experiments. Not only were the scales of these anomalous effects insensitive to intervening distance and time but, like the PRP data, they also displayed similar structural patterns to those of the corresponding local experiments,[3] again suggesting that both the human/machine and PRP anomalies, previously regarded as distinct phenomena, actually might derive from

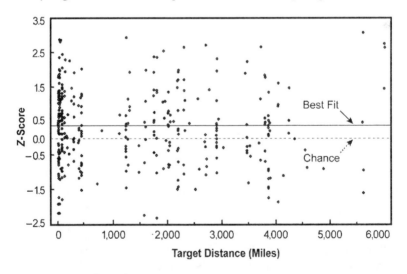

Fig. III-4. 336 binary PRP trial scores as a function of distance.

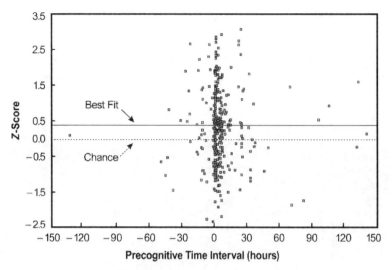

Fig. III-5. 336 binary PRP trial scores as a function of intervening time.

the same non-local mechanism of information exchange. This in turn raises the question of whether space and time are truly intrinsic qualities of the physical world, or whether, as many prominent thinkers have maintained, they are *subjective* coordinates that consciousness imposes in order to organize its experiences. One might go even further, to ask along with Larry Dossey:

> Why speak of the soul in an age of science? Why suggest that there is some aspect of the psyche that is not subject to the limitations of space and time, and which might precede the birth of the body and survive its death? The main reason is that something vital has been left out of almost all the modern efforts to understand our mental life — something that accounts as a first principle, without which everything is bound to be incomplete and off base. This missing element is the mind's nonlocal nature: the soul.[4]

References

[1] Erwin Schrödinger. *What is Life? and Mind and Matter.* Cambridge: The University Press, 1967. p. 133.

[2] Brenda J. Dunne and Robert G. Jahn. "Information and uncertainty: 25 years of remote perception research." *Journal of Scientific Exploration, 17,* No. 2 (2003). pp. 207–241.

[3] Brenda J. Dunne and Robert G. Jahn. "Experiments in remote human/machine interaction." *Journal of Scientific Exploration, 6,* No. 4 (1992). pp. 311–332.

[4] Larry Dossey. *Recovering the Soul: A Scientific and Spiritual Search.* New York: Bantam Books, 1989.

6

WHAT IS A SCIENTIST TO DO?

Inspect every piece of pseudoscience and you will find
a security blanket, a thumb to suck, a skirt to hold.
What does the scientist have to offer in exchange?
Uncertainty! Insecurity!
— Isaac Asimov[1]

The essence of the scientific method is the acquisition of knowledge by observing, measuring, and systematizing the experiences of consciousness within certain conceptual frameworks, including space and time, which are assumed to be objective qualities of the physical world. When faced with phenomena that appear to depart from such metrics, we speak of them as anomalous and react to them with understandable skepticism. But scientific study of anomalous phenomena that involve inescapably subjective features introduces further troublesome aspects: How does one define what is being measured? How does one quantify the immeasurable? How does one control for subjective variables in an experimental design? How can they be conceptualized and represented? And finally: How objective is *any* process of measurement?

What is a scientist to do when confronted with empirical evidence that flies in the face of contemporary scientific belief? According to one Nobel laureate physicist, "The conventional answer would be that this evidence must be tested with an open mind and without theoretical pre-conceptions."

This would appear to be an excellent response; but unfortunately, he then adds "I do not think this is a useful answer, but this view seems to be widespread," and follows this with the less-than-open-minded assertion: "...there is no room in our world for telekinesis or astrology...or other superstitions." (We suspect that he also would have included remote perception in his list of "superstitions.") Nevertheless, he eventually acknowledges that "If it could be shown that there is any truth to any of these notions it would be the discovery of the century, much more exciting that anything going on today in the normal work of physics."[2] He had it right at the beginning and end of his statement. Given the monumental implications of phenomena that appear to depart so radically from traditional physical concepts, it seems not only reasonable, but obligatory to try to understand them through further systematic study.

Our quarter-century-long odyssey to attempt quantification and interpretation of the PRP phenomenon taught us a great deal, including the humility that came with realizing that there was still much more to learn about the mechanics and correlates of this remarkable information modality. Before indulging in the somewhat speculative conjectures that we might credibly offer at this point in our understanding, it might help first to assemble a synoptic reprise of our findings.

The evidence acquired in the early remote perception trials had raised profound questions in the minds of the PEAR researchers, similar, no doubt, to those of the countless others who have experienced first-hand the validity of Paracelsus's extraordinary claim. But it had also persuaded us that the phenomena were real, albeit elusive, and that more rigorous scientific study was merited. Our subsequent efforts to devise strategies capable of representing the amount of anomalous information acquired in this process had followed traditional scientific methods, *e.g.* performing experiments under carefully controlled conditions; systematically eliminating sources of extraneous noise to bring the effects into sharper focus; and posing theoretical models to dialogue with the empirical results.

The first phases of this effort had provided encouraging indications that a set of standardized queries, addressed to both the agent's description of the physical target and to the percipient's stream-of-consciousness narrative, could serve as a form of information net to capture the essence of the anomalous communication. *Ex post facto* application of this technique to existing data confirmed the efficacy of this approach, producing results that were reasonably consistent with previous human judge assessments and encouraging continued applications.

The second phase of the program entailed *ab initio* utilization of this method of analysis on a new body of experiments that also produced highly significant results. Although the average effect size was considerably smaller than that of the original *ex post facto* subset, the statistical yield still was sufficiently robust to indicate that the method continued to serve its intended purpose. Further application on a second segment of *ab initio*–encoded trials, however, compounded to a substantially lower yield, and it became apparent that a major problem was surfacing. Since both phases of the *ab initio* portion of the program utilized identical descriptor questions and scoring algorithms, their analytical effectiveness could not have been the source of the lower yield in the later phases of the program, so some subtler influence was evidently at work.

Like so much of the research in consciousness-related anomalies, further replication, enhancement, and interpretation of these effects proved progressively more elusive. As the analytical techniques became more sophisticated, the empirical results became weaker. It appeared as if each subsequent refinement of the analytical process, intended to improve the quality and reliability of the information net, had resulted in greater reductions of the amount of raw information being captured. This diminution of the experimental yield prompted *ad hoc* examinations of numerous factors that could have contributed to it, but after exploring and precluding various possible sources of statistical or procedural artifacts associated with the descriptors themselves, we were forced to conclude the cause of the problem more likely lay in the subjective sphere of the experience. We recalled that when queried about their personal reactions to the encoding process, participants' most common complaint was a feeling of being "constrained" by the required forced-choice queries. Although the FIDO and distributive methods had been implemented to permit participants more freedom in formulating their responses and had produced a number of impressionistically successful trials, the composite quantitative results they returned were only marginally significant in the former case, and completely consistent with chance in the latter.

These failures to reinvigorate the experimental program made it evident that the efforts to enhance the effect by progressively more elaborate analysis techniques had actually been counterproductive. While the prior methods had served their intended analytical purpose reasonably well, the attenuation of subsequent yields indicated that some kind of interference between the more elaborate analytical measures and the generation of the effects they were attempting to measure was in play. Figure III-6 displays the comparative effect sizes of the data from these various experimental periods, emphasizing the systematic decrease of the yield beginning with the implementation of the *ab initio* binary

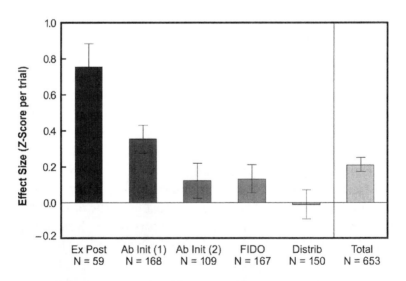

Fig. III-6. **Effect sizes of various PRP data subsets.**

experiments. The corresponding numerical values of these segments are presented in Table III-2, which indicates that while the
composite effect of the overall PRP database remained highly significant, this bottom-line result was driven primarily by the much
stronger yields of the earlier trials, bolstered by the substantial size
of the entire database.

TABLE III-2

PRP Summaries by Database

Database	# Trials	# Series	# Agents	# Percipients	Total # Participants	Composite z-Score	Effect Size	Probability
				# Participants*				
Ex post facto	59	7	4	13	16	5.792	.754	3×10^{-9}
Ab initio	277	42	13	26	30	4.378	.263	6×10^{-6}
Initial Trials	168	29	9	21	23	4.582	.354	2×10^{-6}
Later Trials	109	13	7	13	15	1.291	.124	.098
FIDO	167	9	19	22	25	1.735	.134	.041
Distributive	150	30	15	15	16	−0.108	−.009	.543
TOTAL	653	88	39	59	69	5.418	.212	3×10^{-8}

*Some participants contributed to more than one database, in both agent and percipient capacity.

Returning to the raw free-response data, we became aware of
another indicative pattern, *i.e.* the written free-response perceptions in the later trials were considerably shorter than those generated in the earlier ones, some of which had run to several pages
of narrative. Indeed, in many of the later trials, percipients' verbal
descriptions consisted of only a few cursory phrases, apparently
intended simply to clarify nuances of their descriptor responses,
and provided little in the way of the stream-of-consciousness
imagery they had been asked to record. It appeared that as the
percipients became more familiar with the descriptor questions,
their subjective impressions were increasingly guided and circumscribed by them, as though the questions were establishing the informational framework for their responses. Thus, the experiment
had ultimately taken on the characteristics of a multiple-choice

task, and the locus of the experience had shifted from the realm of intuition to that of intellect.

References

[1] Isaac Asimov. *Past, Present, and Future.* Buffalo, NY: Prometheus Books, 1987. p. 65.

[2] Steven Weinberg. *Dreams of a Final Theory.* New York: Pantheon Books, 1992. pp. 48–50.

7

FROM ANALYSIS TO ANALOGY

Analogies, it is true, decide nothing
but they can make one feel more at home.
— Sigmund Freud[1]

Having exhausted the search for credible causes of the remote perception signal deterioration within the analytical techniques themselves, we turned to more subtle subjective aspects of the protocol for hints of possible sources of the problem. From this perspective, we noted that all of the methodological refinements in the scoring techniques had been directed either toward more efficient extraction of the anomalous information, or to reduction of possible sources of ambiguity, artifact, or bias. Some had been efforts to "sharpen" the remote perception "signal;" others were attempts to "tighten" the experimental "controls;" and a few were designed to "clarify" certain characteristics of the communication "channel." All of these terms reflected an emphasis on achieving increasingly precise specification and reducing the noise or uncertainty in the process. Yet each increment of these analytical refinements appears to have resulted in a systematic reduction not of the "noise," but of the "signal" itself, raising the radical possibility that manifestation of the anomaly might actually require a certain degree of the very noise, or uncertainty, that we had invested so much effort to reduce! It was a possibility, however, for which precedent could be found in other domains of scholarly inquiry, and therefore seemed worth consideration in this context, as well.

The most immediate technical examples of this complementarity of signal and noise also could be found in our own human/machine experiments. All of these studies employed some form of random processor, and the anomalous effects manifested as departures of their statistical outputs from chance expectations. It was as if the random noise provided the essential raw material out of which the mind of the operator was somehow able to construct a small amount of ordered signal.

But such effects are by no means restricted to anomalies research, *per se*. Similar departures from canonical expectations can be found in contemporary engineering applications known as "stochastic resonance," wherein a deliberate increase in the overall level of noise in certain kinds of lasers or sensitive electronic circuits appears to enhance the detection of weak, fluctuating signals.[2,3] Other studies have demonstrated that the introduction of an element of chaos into various types of nonlinear processes, such as the interaction of two otherwise independent random oscillators, could stimulate synchronous behavior between the transmitter and the receiver.[4,5] In either of these instances, information or order was being introduced into a sensitive nonlinear physical system, not by reducing the ambient noise, but by *increasing* it.

Of particular relevance for our purpose were unanticipated corollary observations that in such synchronization processes the receiver could record changes in the signal *before* the transmitter registered the transmission of those changes. In other words, these systems seemed capable of anticipating the synchronization. The engineers who carried out one of these studies remarked:

> We would thus expect that any of those analogous systems which exhibit chaos should also be liable to anticipating synchronization. We thus hope that our work will act as a stimulus to explore the opportunities for observing anticipating synchronization in physical, chemical, biological and socioeconomic systems.[3]

Following this proposition, we pondered whether in a certain sense the remote perception process qualified as an example of a "sensitive nonlinear system with a weak fluctuating signal" that entailed a certain degree of chaos, and that the participants in these experiments were functioning as "two otherwise independent random oscillators." Could it be that our PRP information was also dependent upon a background of random noise for its manifestation? If so, it would appear that our attempts to enhance the signal by sharpening the information channel could, in fact, have been at least partially responsible for its attenuation. We might also note in passing the similarity between the anticipatory capability of these physical systems and results reported in the parapsychological literature from certain bodies of research on so-called "presentiment" phenomena.[6]

Reaching beyond the physical sciences for relevant analogies, recall that the accepted model of biological evolution emphasizes the importance of uncertainty in enhancing cellular level information (*cf.* Chapter I-3). Darwinian theory posits that living species adapt to their environment by selecting for specific beneficial traits that emerge in the process of *random* genetic mutations. This process is itself strongly dependent on the reservoir of biological "noise" resident in the massive redundancy of continuously recombined genetic information. When the randomness of this process is limited, as in repeated interbreeding, the short-term advantage of increased predictability of inherited traits is offset by longer-term weakening of the genetic strain of the species. And, consistent with the Neo-Lamarckian hypothesis, it is not too unreasonable to contemplate the possibility that individual or collective consciousness of sentient beings might be capable of introducing order into this evolutionary random process via their intention and resonance with it.

Insights could also be derived from quite a different realm of human experience, namely, the practice of certain mystical divinatory traditions where anomalous relationships between signal and noise are also exploited. In most of these, a clearly defined question is submitted to some kind of *random* process in an effort to access information unavailable to the conscious mind. Typically, the response comes in imprecise or symbolic form that requires translation into meaningful or pertinent terms. One well-known example is the renowned ancient Greek Oracle of Apollo at Delphi, a highly respected source of wisdom that played a central role in classical Greek culture and politics. Consultation of the oracle involved a priestess called the Pythia who, crowned with laurel and reputedly in an altered state of consciousness stimulated by vapors arising from a cleft in the earth over which she sat on a tripod, produced a "free response" utterance that was then interpreted by an attending priest in response to the seeker's query. Two points of potential relevance here are the non-analytical, receptive state of mind of the "percipient," and the deferment of interpretation by the "judge" until *after* the experience had been completed.

Another ancient oracle, still widely in use, is the Chinese "Book of Changes," or *I Ching*, a divination process that involves generation of a sequence of random binary events (originally the casting of a handful of yarrow sticks, and later the tossing of coins), the results of which are represented as two "trigrams." These are referred to a table, or matrix, that identifies each of 64 possible binary combinations, or "hexagrams," with a specific text that is then consulted to obtain a response to the original query — again *after* the hexagram has been identified. Notwithstanding the subjective nature of the text interpretation, a vast body of evidence accumulated over many millennia testifies to the efficacy of the *I Ching* in producing accurate and consequential information. Despite the claim of many rationalists that such oracles are simply bizarre combinations of sufficiently ambiguous wording and wishful thinking, this formula is similar to that underlying the remote perception process that has now been demonstrated by rigorous analytical quantification to convey more meaningful information than can be attributed to "mere chance."

Psychologist Carl Jung, who devoted more than 30 years to the study of the *I Ching*, in his Foreword to the classic Richard Wilhelm translation observed that

> ... we know now that what we term natural laws are merely statistical truths and thus must necessarily allow for exceptions

... If we leave things to nature, we see a very different picture: every process is partially or totally interfered with by chance, so much so that under natural circumstances a course of events absolutely conforming to specific laws is almost an exception.[7]

Jung also noted the emphasis placed by the ancient Chinese mind on chance, and the subjective interpretation of events in the modern world of quantum mechanics, where the behavior of inherently random microscopic physical events includes the observer as well as the observed. In both domains, what Jung referred to in his Foreword as the "hidden individual quality of things and men"[6] draws on the unconscious and intangible attributes that undergird both the experiences of the conscious mind and the behavior of the tangible physical world, respectively, in a fashion similar to the conceptual framework utilized in our "Modular Model of Mind/Matter Manifestation (M^5),"[8] to be reviewed in the next Section. Both Jung's representation and our own emphasize that the causal and synchronistic perspectives of reality are *complementary*, rather than mutually exclusive. Jung maintained that the "coincidence" of a synchronistic event occurs "because the physical events are of the same quality as the psychic events and because all are the exponents of one and the same momentary situation."[6] Our M^5 representation of this concept speaks of the emergence of both cognitive experience and physical events from a common underlying substrate where the unconscious mind and the undifferentiated world of physical potentiality merge, and where the distinction between mind and matter blurs into uncertainty. Given this common origin, it should not be surprising to observe correlations between their manifested expressions in the worlds of both mental and physical "reality." And just as the concept of complementarity in quantum mechanics brings with it an inescapable degree of *uncertainty* that makes it impossible to achieve absolute precision in the two frames of reference simultaneously, the complementarity of an "objective" causal picture of reality and a "subjective" synchronistic one also may necessitate tolerance of a degree of uncertainty in both respects.

In many ways, the empirical evidence from our remote perception and other forms of anomalies research is equally compatible with an acausal, or synchronistic, model as with a causal one. Although we had recognized this in principle, our experimental approach and the language we deployed in describing the effects nevertheless betrayed a dominance of intrinsically causal assumptions. For example, despite repeated comments from participants that the PRP experience felt more like "sharing" than "sending and receiving," we persisted in speaking of information "transmission." Even our enduring efforts to extract the "signal" from the "noise" reflected a more deterministic orientation. Yet, Jung's model, the ancient divinatory traditions, evolutionary theory, contemporary signal processing research, and our own human/machine anomalies data and models all suggest that noise may be a requisite component of the creation of information, and that totally objective linear causality may not obtain under these circumstances.

If one identifies "noise" in the remote perception context with the percipient's uncertainty, or lack of conscious knowledge, about the target, and "signal" as the content of valid information acquired in the process, the early experiments, wherein percipients were asked simply to generate an unfocused, free-response stream of consciousness, were in this sense more "noisy" than the later efforts where percipients' imagery was guided by a more structured information "grid" or "filter" of descriptor queries. In those trials that were only encoded *ex post facto*, the participants had no knowledge of the information filter that would subsequently be imposed after the data were generated, and they seemed more easily able to access information about the targets. In the first generation of *ab initio* binary-encoded trials, when descriptor check-sheets were something of a novelty and percipients were still urged to generate their free-response descriptions before attempting descriptor encoding, the transcripts tended to be somewhat shorter, but most still comprised a free-association type of narrative. These trials also produced successful results, albeit of a somewhat smaller average effect size.

In the later *ab initio* experiments, however, the greater confidence that had been acquired in the efficacy of the analytical judging approach led to less importance being placed on the raw free-response data, by both participants and experimenters, and this shift of emphasis was reflected in abbreviated, even cursory, percipient responses. In retrospect, it is apparent from the content of these shorter transcripts that the percipients were anticipating the descriptor questions and inadvertently focusing their attention on those particular aspects of their experience. Although the goal of the quaternary FIDO, and then the distributive descriptor questions, was to relieve the participants' reported feelings of constraint, these more complex forms of questions appear to have had just the opposite effect, forcing percipients to pay even more attention to the nuances of the information grid, thereby screening out any portions of the signal that were not apparently relevant to it. In this way, the background "noise" was further impoverished, and more cognitive processes, associated with achieving internal consistency in what essentially had become a forced-choice task, effectively restricted the flow of unconscious imagery.

It is also telling that for several years this trend had not been perceived as a problem by our researchers. Typing 30 numbers into a computer was much easier than evaluating lengthy verbal transcripts, and the efficiency of more rapidly acquiring a quantitative indication of the merit of an individual trial increasingly replaced the spontaneous excitement of finding apparent correspondences in the raw data. The shift in experimental perspective from predominantly subjective to almost totally analytical was so gradual that little consideration was given to the possible costs of such a transition. For example, combination of the data from the first and second phases of the *ab initio* experiments was justified solely on technical grounds. No serious consideration was given to the subjective implications of shifting from ranking the overall impressionistic quality of a trial to measuring its specific information content, other than the relative efficiency and statistical power of the two approaches.

The larger effect size of the instructed vs. the volitional trials also supports the importance of retaining an adequate component of noise or uncertainty in the system. When percipients attempted to describe scenes chosen by a random process that precluded utilization of prior knowledge about the agent's habits or personal preferences, their perceptions displayed a larger component of anomalous information. In the volitional protocol, where one might imagine a certain *a priori* advantage based on the percipient's familiarity with the agent's preferences or behavior patterns, such rational expectations may have imposed yet another kind of information filter that inhibited the subtle information and detection processes. Again, the stronger "signals" appear to have been generated under the "noisier" conditions, *i.e.* in the absence or minimization of any orderly or rational form of imagined constraint.

In yet more comprehensive reflection, we could note that virtually any form of creative accomplishment proceeds from an initial conceptual vision to some tangible construction, enabled by the particular talents, tools, and determination of the creator, and the resulting products thereof serve to convey both the original inspiration and its particular purposeful implementation to an audience. Such concept/construct transpositions are at work within this remote perception phenomenon as well: an agent positioned in an explicit physical scene allows its details to engender a corresponding but more diffuse pattern of generic sensations, which are, by whatever process, more amenable to being shared with the psyche of the percipient, who then inverts the transform to create from these impressionistic rudiments his or her own version of a scene consistent with them. No wonder that some details are lost or misplaced, or that spurious elements are added in this second conversion process, or that time and space become less salient correlates! A different set of memory equipment and processes must be involved in the information acquisition phase, dreamlike in nature and therefore more "wavelike" in character, while its reconstruction to a more explicit "particulate" description must continue to rely upon more conventional cognitive assembly.

It is even possible that the overall success of these experiments derives in considerable measure from the irrational nature of the remote perception task itself. When requested to describe a spatially and temporally remote scene without access to any known sensory channel, percipients are forced to abandon any rational strategy for fulfilling such an assignment. With cognitive functioning thus confounded by uncertainty, leaving the conscious mind less able to mask the subtle signal with rational associations, the unconscious mind of the percipient may better be able to access the "hidden individual quality of things and men."

Although a degree of uncertainty may indeed be necessary for the generation of remote perception effects, the complementary relationship between signal and noise nevertheless requires retention of a certain amount of contextual structure in the process. For example, in some of the exploratory trials where percipients did not know the identity of the agent or the time of target visitation, the results were virtually null, completely consistent with chance expectations. As in the *I Ching* or other divinatory arts, where it is essential that the querent pose a clearly defined question, the remote perception process also seems to require the percipient to establish some minimal frame of reference when addressing the unknown target. If indeed such a process involves an excursion into the unconscious realm of undifferentiated potential in order to acquire specific information, some corresponding specific question would appear to be prerequisite.

If the study of remote perception is to qualify as a scientific enterprise, some form of quantitative assessment of the amount of anomalous information acquired is inescapable. While the emergence of consciousness-related anomalous physical effects may indeed be driven by a host of subjective factors, our efforts to demonstrate, record, and quantify them necessarily entail the imposition of objective criteria and measurements. The challenge is that the former appears to be obstructed by the latter, and vice versa, leaving us to find a way to straddle the subjective/objective dichotomy with some optimized compromise. In our case, while our efforts to establish defensible and quantitative analytical techniques have helped to validate the remote perception phenomenon, carried to excess they appear to have progressively suffocated its emergence.

Whether this interference functions primarily in the psyches of the human participants, or is more endemic in the physical characteristics of the information itself, is unclear and possibly irresolvable. Notwithstanding, similar indications have emerged from a number of our other experiments, collectively suggesting that this uncertainty is not merely a limitation on the attainable empirical precision, but is evidence of the fundamental importance of "noise" as the raw material out of which the anomalous information is constructed.

To this end, we have proposed that a more astute balance between the analytical and aesthetic dimensions of such phenomena needs to guide future explorations of consciousness-related anomalies.[9–12] In an article entitled "Science of the Subjective,"[13] we observed how in the interplay of objective intellect and subjective spirit, we are dealing with the primordial conjugate perspectives whereby consciousness triangulates its experience. Consistent with the adage that the whole is greater than the sum of its parts, the disproportionate focus in our remote perception experiments on the descriptor components, rather than on the gestalt of the phenomenon, seems to have resulted in obscuring the essence of the whole. It now appears that the subjective quality of these experiences is more effectively enhanced when less encumbered by analytical or cognitive overlays, and when their inherent uncertainties are both acknowledged and utilized. This was borne out in a modest exploratory exercise we conducted wherein an agent constructed artificial targets from sequences of random numbers that were arbitrarily associated with the list of descriptors. These "pseudo targets" turned out to be rather unrealistic, yet in over half of the trials a percipient was able to construct narratives that included a majority of correct descriptors. Unfortunately, we did not have an opportunity to pursue such an exercise more systematically, but a return to emphasis on free-response descriptions could be a more productive strategy for any future experiments. This analytical/aesthetic paradox was summarized quite succinctly by Larry Dossey in his book *The Power of Premonitions*:

> When we put [psi experiences] under the microscope, we hinder their appearance.... This doesn't mean they don't exist, or that the prior experiments documenting their existence lied, but that we have hounded them into hiding with our petulant demands.[14]

References

[1] Sigmund Freud. New Introductory Lectures on Psycho-Analysis (1933). In James Strachey, Ed. *The Standard Edition of the Complete Psychological Works of Sigmund Freud* (1964), Vol. 22. p. 72.

[2] Bruce McNamara, Kurt Wiesenfeld, and Rajarshi Roy. "Observation of stochastic resonance in a ring laser." *Physical Review Letters, 60*, No. 25 (1988). pp. 2626–2629.

[3] Bruce McNamara and Kurt Wiesenfeld. "Theory of stochastic resonance." *Physical Review A, 39*, No. 9 (1989). pp. 4854–4869.

[4] R. J. Jones, P. Rees, P. S. Spencer, and K. Alan Shore. "Chaos and synchronisation of self-pulsating laser diodes." *Journal of the Optical Society of America, B, 18*, No. 2 (2001). pp. 166–172.

[5] S. Sivaprakasam, E. M. Shahverdiev, P. S. Spencer, and K. Alan Shore. "Experimental demonstration of anticipating synchronization in chaotic semiconductor lasers with optical feedback." *Physical Review Letters*, *87*, No. 15 (2001). 154101.

[6] Dean I. Radin. *Entangled Minds: Extrasensory Experiences in a Quantum Reality*. New York: Simon & Schuster, 2006.

[7] Carl Gustaf Jung. Foreword to *The I Ching* (R. Wilhelm, trans.). Princeton, NJ: Princeton University Press, 1950. pp. *xxxi–xxxix*.

[8] Robert G. Jahn and Brenda J. Dunne. "A modular model of mind/matter manifestations (M^5)." *Journal of Scientific Exploration*, *15*, No. 3 (2001). pp. 299–329.

[9] Robert G. Jahn. "Anomalies: Analysis and aesthetics." *Journal of Scientific Exploration*, *3*, No 1 (1989). pp. 15–26.

[10] Robert G. Jahn. "The Complementarity of Consciousness." In K. R. Rao, ed., *Cultivating Consciousness for Enhancing Human Potential, Wellness, and Healing*. Westport, CT and London, UK: Praeger, 1993. pp. 149–163. (Also available as Technical Note PEAR 91006. Princeton Engineering Anomalies Research, Princeton University, School of Engineering/Applied Science. December 1991.)

[11] Brenda J. Dunne. "Subjectivity and Intuition in the Scientific Method." In R. Davis-Floyd and P. Sven Arvidson, eds., *Intuition: The Inside Story*. New York and London: Routledge, 1997. pp. 121–128.

[12] Robert G. Jahn. "M*: Vector representation of the subliminal seed regime of M^5." *Journal of Scientific Exploration*, *16*, No. 3 (2002). pp. 341–357.

[13] Robert G. Jahn and Brenda J. Dunne. "Science of the subjective." *Journal of Scientific Exploration*, *11*, No. 2 (1997). pp. 201–224.

[14] Larry Dossey. *The Power of Premonitions: How Knowing the Future Can Shape Our Lives*. New York: Dutton (The Penguin Group), 2009. p. 138.

SECTION IV
Thinking Outside the Box

1

ACCOMMODATING ANOMALIES

*There is another great and powerful cause why
the sciences have made but little progress;
which is this. It is not possible to run a course aright
when the goal itself has not been rightly placed.*
— Francis Bacon[1]

The essence of the scientific method as originally proposed in 1620 by Sir Francis Bacon is an ongoing dialog between experiment and theory, a process we have whimsically re-labeled the "scientific two-step."[2] In its most common format, empirical studies provide observations and measurements that stimulate explanatory models, which in turn suggest refinements for the next round of experiments. Such has been the case in such areas of physical science as combustion research, turbulent fluid dynamics, and complex and chaotic systems dynamics, to name but a few where empirical results have tended to lead theoretical representation. But in other cases, new theoretical concepts have sought verification by dedicated empirical efforts, as in the celebrated confirmation of Einstein's general relativity theory via observations on the perihelion of the planet Mercury or, more recently, by experimental confirmations

© 2007 Viṣudi
www.Visudi.com

A self-organizing vortex

of Bell's paradox and quantum entanglement. A similar process has prevailed in many other scientific areas as well.

In either sequence, if the experimental and theoretical "feet" are both sound, well balanced, and working in concert, science can move productively forward, shifting its weight successively from empirical experiment to explicating theory or from predictive theory to confirmatory experiment. But when the data are inconsistent with the model, or when the theoretical predictions conflict with well-established empirical data, the scientist is confronted by an anomaly that signals that either the observations are artifacts of flawed experimentation, or that the theory is inadequate to accommodate the full weight of the observed phenomena. Proper assessment and assimilation of such anomalies require technical skill and experience, complemented by intuition, imagination, inspiration, and a willingness to concede ignorance, in order to identify a more creative path along which the science can resume more fruitful bipedal progress.

The anomalous empirical findings of the PEAR laboratory have presented a bewildering array of irregularities, contradictions, and departures from canonical precedents and expectations that essentially denude conventional modeling strategies of any hope of effectiveness. Given the considerable effort expended to guarantee that these results cannot be dismissed as experimental artifacts, their stark inconsistencies with established physical and psychological presumptions require the development of more competent new theoretical models that are capable of accommodating these strange effects. In particular, it has become clear that any posited models must deal with a hierarchy of extraordinary features:

- Tiny informational increments riding on random statistical backgrounds;
- Correlations of objective physical evidence with subjective psychological parameters, most notably intention, attitude, meaning, resonance, and uncertainty;

- Independence of the character or magnitude of the effects on intervening distance and time;
- Oscillatory sequential patterns of anomalous performance;
- Data distribution structures consistent with alterations in the prevailing elemental probabilities;
- Complex and irregular replicability.

These exceptional empirical characteristics necessarily have turned us toward similarly radical theoretical propositions, the overarching character of which has been the involvement of a proactive, subjective consciousness, as specified in an article entitled "Science of the Subjective"[3] in the following terms:

Any disciplined re-admission of subjective elements into rigorous scientific methodology will hinge on the precision with which they can be defined, measured, and represented, and on the resilience of established scientific techniques to their inclusion. For example, any neo-subjective science, while retaining the logical rigor, empirical/theoretical dialogue, and cultural purpose of its rigidly objective predecessor, would have the following requirements: acknowledgment of a proactive role for human consciousness; more explicit and profound use of interdisciplinary metaphors; more generous interpretations of measurability, replicability, and resonance; a reduction of ontological aspirations; and an overarching teleological causality. More importantly, the subjective and objective aspects of this holistic science would have to stand in mutually respectful and constructive complementarity to one another if the composite discipline were to fulfill itself and its role in society.

Within this overarching framework, our particular efforts thus far have explored three conceptual models:

1) On the Quantum Mechanics of Consciousness, with Application to Anomalous Phenomena,[4,5]

2) A Modular Model of Mind/Matter Manifestations (M^5),[6,7]

3) Sensors, Filters, and the Source of Reality.[8,9]

Each of these has been developed in considerable technical detail in their respective references, but they will be recast and summarized in the following chapters. In considering these presentations, we recommend that the reader not attend so much to the apparent distinctions among them as to the ensemble of common features they embody, for therein lies our best hope of identifying and clarifying salient aspects of the Source of reality itself and of our relationship to it.

Collectively these efforts comprise rather primitive attempts to erect a skeletonic framework on which a more comprehensive Science of the Subjective may eventually be supported, a daunting aspiration that will only be consummated after the prevailing rules of scientific engagement have submitted to major modifications.

References

[1] Francis Bacon. Translation of *Novum Organum*, LXXXI. In James Spedding, *The Works of Francis Bacon* (1864), Vol. 8. p. 113.

[2] Robert G. Jahn and Brenda J. Dunne. *Margins of Reality: The Role of Consciousness in the Physical World*. San Diego, New York, London: Harcourt Brace Jovanovich, 1987. Reprinted: Princeton, NJ: ICRL Press, 2009. (Section I, Chapter 1, "Scientific Two-Step," pp. 3–9.)

[3] Robert G. Jahn and Brenda J. Dunne. "Science of the subjective." *Journal of Scientific Exploration*, *11*, No. 2 (1997). pp. 201–224.

[4] Robert G. Jahn and Brenda J. Dunne. "On the quantum mechanics of consciousness, with application to anomalous phenomena." *Foundations of Physics*, *16*, No. 8 (1986). pp. 721–772. (An Appendix [in the form of a Technical Note] is also available, which contains a collection of relevant quotations by many of the patriarchs of modern physics.)

[5] Jahn & Dunne. *Margins of Reality*. (Section IV; The Waves of Consciousness, pp. 195–287.)

[6] Robert G. Jahn and Brenda J. Dunne. "A modular model of mind/matter manifestations (M^5)." *Journal of Scientific Exploration*, *15*, No. 3 (2001). pp. 299–329.

[7] Robert G. Jahn. "M*: Vector representation of the subliminal seed regime of M^5." *Journal of Scientific Exploration*, *16*, No. 3 (2002). pp. 341–357.

[8] Robert G. Jahn and Brenda J. Dunne. "Sensors, filters, and the Source of reality." *Journal of Scientific Exploration*, *18*, No. 4 (2004). pp. 547–570.

[9] Zachary C. Jones, Brenda J. Dunne, Elissa S. Hoeger, and Robert G. Jahn, Eds. *Filters and Reflections: Perspectives on Reality*. Princeton, NJ: ICRL Press, 2009.

2

CONSCIOUSNESS AND
QUANTUM MECHANICS

*All ideas we form of the outer world are ultimately
only reflections of our own perceptions. ... Are not all
so-called natural laws really nothing more or less
than expedient rules with which we associate the run
of our perceptions as exactly and conveniently as possible?*
— Max Planck[1]

Quite early in the program we were struck by a number of similarities between the historical and philosophical evolution of quantum science and the ongoing unfolding of the experience and representation of consciousness-related physical anomalies.[2] In both scenarios, classically respected conceptual and analytical models of reality were challenged by the advent of diverse bodies of anomalous observations made possible by the development of more sensitive and reliable experimental equipment. In each case, extensive attempts to rationalize the empirical effects within prevailing formalisms proved categorically and profoundly unsuccessful, forcing postulation and development of a number of counter-intuitive propositions. Among those originally posed in the context of atomic-scale physical events were the quantization of energy and other physical observables; the probabilistic character of quantum observations; the wave/particle duality and the wave mechanics of atomic structure; and the uncertainty, complementarity, exclusion, and indistinguishability principles.[3] Many of these physical anomalies, especially the apparent role of the observer in the establishment of atomic-scale reality, appear to bear striking metaphorical resemblance to our own consciousness-related anomalies, suggesting that all of these effects might better be regarded as impositions by the experiencing consciousness, rather than as intrinsic characteristics of the physical events, *per se.*[2,4,5]

From this perspective, all of reality may be viewed as the product of a holistic and fully complementary dialog between consciousness and the environment in which it is immersed, the vocabulary of which is information. For our purposes, the definition of "consciousness" is intended to subsume all categories of human experience processing, including perception, cognition, intuition, instinct, and emotion, at all levels, including those commonly termed conscious, subconscious, or unconscious, and presumes no specific psychological or physiological mechanisms. Our concept of "environment" includes all circumstances and influences affecting the consciousness that are perceived by it to be separate from itself, including, under some circumstances, its own physical corpus. Thus, consciousness and environment are represented as engaging in an ongoing "I/Not I" dialogue, with the specification of the interface between the two necessarily subjective and situation-specific. The term "reality" is invoked in an existential sense, subsuming "observation," "experience," and "behavior."

With these definitions in hand, the model proceeds from the following premises:

1) The purpose of any theory, or of any other scheme of conceptual organization, is to order and correlate the experiences of consciousness interacting with its environment. Neither that environment, nor that consciousness, can properly be addressed in isolation; only in the interaction of the two is reality constituted.

2) The primary currency of reality is information, which may flow in either direction; *i.e.*, consciousness may insert information into the environment as well as extract information from it.

3) The common ingredients of physical theories, such as mass, momentum, and energy; electric charges and magnetic fields; the quantum and the wave function; and even distance and time; are no more than useful organizing concepts that consciousness has

conceived and utilized to enable it to order the blizzard of information bombarding it from its environment, or passing from it to its environment. As such, these concepts reflect as much the characteristics of consciousness as those of the environment. More precisely, they reflect the characteristics of consciousness interacting with its environment.

4) It follows, then, that the models, imagery, and formalisms of any physical theory may have metaphoric utility for the representation of the nature and processes of consciousness, including those of consciousness examining itself. Conversely, the impressionistic descriptors of subjective experience may, in some form, be requisite ingredients of any truly general theory of reality.

In this spirit, the features and formalisms of rudimentary quantum mechanics can be appropriated via suitable metaphors to represent the characteristics of these consciousness/environment interactions. For example, as an alternative to the prevailing "particulate" view of consciousness, wherein an individual consciousness is typically presumed to be well localized in space and time and to interact only with a few proximate aspects of its environment, we propose that it also be permitted a complementary wave-like conceptualization. This allows us to model it in terms of a quantum mechanical wave function, and its environment, including its own physical corpus, by an appropriate potential energy profile. It then becomes possible to apply any quantum formalism, *e.g.* the Schrödinger wave mechanics, to define eigenfunctions and eigenvalues that can be associated with the cognitive and emotional experiences of that consciousness in that environment (Figure IV-1).

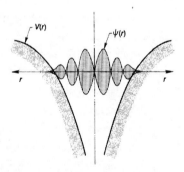

Fig. IV-1. Consciousness eigenfunctions, $\psi(r)$, in an environmental well, $V(r)$[2,4].

To articulate this metaphor more fully, we also may associate certain quantitative aspects of the quantum formalism, such as the coordinate system and its metric, or the quantum numbers of the discrete states, with various qualitative descriptors of consciousness, such as its intensity, perspective, approach/avoidance attitude, balance between cognitive and emotional activity, and "yin/yang" or receptive/active disposition. With these in hand, appropriate computational applications can display metaphoric relevance to individual and collective experiences and, in particular, to our experimental situations. For example, such traditional quantum theoretic exercises as the central force field and atomic structure, covalent molecular bonds, barrier penetration, and quantum statistical collective behavior become useful analogies for the representation and correlation of a variety of consciousness states and experiences, both common and anomalous, and for the design and interpretation of experiments to study these systematically. Within these suggestive associations, many "anomalous" consciousness capabilities follow naturally from its own "wave/particle" complementarity; the empirical "resonance" factor can be represented as a form of "molecular bonding"; gender and co-operator effects emerge as a consequence of "electronic spins" and their pairing; FieldREG effects simulate collective particle behavior in various quantum environments; and the conditional replicability features follow from the intrinsic statistical uncertainties of all quantum phenomena.

As just one illustration of this approach, we may postulate "consciousness atoms,"[5] whose properties are analogous to those of their physical counterparts; that is, they embody a wave/particle complementarity, and are assigned quantum numbers that relate to the arrangement of their probability-of-experience patterns. Following this representation, we can further propose the existence of bonded "consciousness molecules," akin metaphorically to the well-known Heitler-London "exchange force,"[6] wherein the characteristics of the original individual atoms are sacrificed in favor of those of the bonded molecular systems. Such bonds are essential to virtually all material structures in the physical domain, including the DNA molecules that establish the characteristics of all living systems.

Clearly this molecular bond metaphor is a long way from quantitative utility in any consciousness context, pending more explicit identification of the wave functions, mathematical operators, and coordinate systems in which they are expressed, but at least one qualitative guideline for the generation and amplification of anomalous consciousness effects is already suggested. Namely, since the relative magnitude of the alterations of the atomic eigenvalues in their molecular configurations depends on the degree of interaction of the participating elements, *i.e.* on the mutual deformation of their separate potential profiles and on the resonance of their individual wave functions, larger consciousness effects should be expected among persons naturally disposed to some merging of their individual identities. Common examples would be biologically driven parent/child or identical twin bonds, but less biological resonances may also qualify to various degrees, as among lovers, close friends, colleagues, or teammates. The model also allows the external environment to play some role in facilitating such resonance, *e.g.* a romantic setting, a comfortable office, a spellbinding theatrical performance, an emotionally charged sporting event, an inspirational vista, or the ambience-enhancing features that we attempted to incorporate in the PEAR laboratory or that prevailed in the successful FieldREG studies. Many aspects of artistic and intellectual creativity display similar mechanics of resonance and indistinguishability, either among the participants, or with their trade tools.

In general, whenever any device, process, or task is perceived to take on meaningful anthropomorphic qualities it acquires thereby the requisite properties of a consciousness wave function for purposes of resonant interaction with its perceiver. Various more elaborate interactions can be postulated as well, perhaps analogous to inorganic quantum mechanical solid-state systems, such as crystals, metals, and semiconductors, wherein either an organized or stochastic behavior of a large assembly of interacting elements subsumes individual properties within a composite wave function of the ensemble.

Crystal structure

The model also explores the implications of other familiar quantum mechanical principles, such as those of indistinguishability, exclusion, correspondence, complementarity, and uncertainty, all of which provide useful metaphors for representing subjective human experience and its mechanics. This should not be totally surprising, since all these concepts, along with those of distance, time, mass, and charge, and their respective dynamical derivatives, were originally appropriated from more common usage during the development of the "exact" sciences, using the more familiar cognitive characteristics and experiences as similes for representing and communicating observations of the physical environment.

In the broadest sense, such metaphoric representations of consciousness need not even be restricted to physical, or indeed to general scientific vocabulary, but could take their analogies from any province that the human mind has attempted to organize or correlate, be it technical or aesthetic, specialized or generic, deductive or intuitive, including all the venues specified as motivating vectors in Section I. For it is precisely through such processes of organization and correlation that the characteristics of reality and of its embracing consciousness are specified, and possibly even manifested.

References

[1] Max Karl Ernst Ludwig Planck. *A Survey of Physical Theory.* Tr. R. Jones and D. H. Williams. New York: Dover Publications, 1960. p. 53.

[2] Robert G. Jahn and Brenda J. Dunne. "On the Quantum Mechanics of Consciousness, with Application to Anomalous Phenomena." Technical Note PEAR 83005. Princeton Engineering Anomalies Research, Princeton University, School of Engineering/Applied Science. December 1983, revised June 1984.

[3] Banesh Hoffmann. *The Strange Story of the Quantum, An Account for the General Reader of the Growth of the Ideas Underlying Our Present Atomic Knowledge.* 2nd ed. New York: Dover, 1959.

[4] Robert G. Jahn and Brenda J. Dunne. *Margins of Reality: The Role of Consciousness in the Physical World.* San Diego, New York, London: Harcourt Brace Jovanovich, 1987. Reprinted: Princeton, NJ: ICRL Press, 2009. Section IV.

[5] Robert G. Jahn and Brenda J. Dunne. "On the quantum mechanics of consciousness, with application to anomalous phenomena." *Foundations of Physics, 16,* No. 8 (1986). pp. 721–772.

[6] Walter Heitler. *Elementary Wave Mechanics.* Oxford: Clarendon Press, 1946.

3

A MODULAR MODEL

Truth dwells in the deeps.
— Friedrich Schiller[1]

While our experimental data have provided ample empirical evidence for the efficacy of unconscious or "non-conscious" intentionality, the lack of a supporting theoretical framework makes it a problematic concept in most scientific circles. Much of the difficulty traces to the prevailing assumption that any "real" phenomenon must obey known physical laws and be explainable in terms of known physical mechanisms. In stark contradiction to the extensive and popular efforts to explicate consciousness in terms of the physical brain from which it is presumed to emanate, the evidence from these PEAR experiments, particularly their non-local correlations, refuses to accommodate itself to such presumptions. Consequently, the Cartesian firewall that has been imposed between mental and physical phenomena remains conceptually impenetrable, and any effort to detect and specify an agency whereby these two domains can communicate directly is precluded. Nevertheless, to paraphrase Galileo, *"Eppure comunicano"* ("And yet they communicate").

In an attempt to circumvent this difficulty without violating the Cartesian imperative, we have articulated a second class of model labeled "M^5: A Modular Model of Mind/Matter Manifestations."[2] In brief, this model postulates that anomalous effects such as those observed in our experiments do not emerge from direct intercourse between the conscious mind and the tangible physical world, but have their origins in the depths of the unconscious mind and in an intangible substrate of physical reality, at whose interface the distinction between mind and matter blurs and each surrenders its independent utility. This is a misty domain of intrinsic uncertainty and conditional probability, where space and time have yet to be defined, let alone distinguished, and where information waits to be born. The taxonomy of this model can be represented by the "duplex" geometry sketched in Figure IV-2.

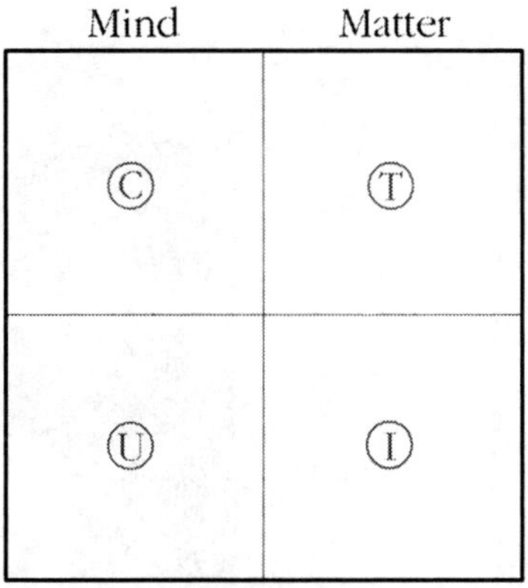

Fig. IV-2. Conceptual geometry of M⁵ model.

The upper left module labeled Ⓒ encompasses all of the commonly recognized capabilities of the conscious mind, subsuming the psychological functions of perception, awareness, cognition, contemplation, representation, organization, memory, volition, activation, etc. The primary processes executed in this domain are the subjective reactions to objective experiences derived from interactions with the tangible physical world and with other consciousnesses, and the logical organization thereof via internal ruminations. In short, this module comprises "Psychology 101." The adjacent upper right module, labeled Ⓣ, subsumes all of the known material substances and structures, dynamic and energetic processes, and information transfer mechanisms of the tangible physical world, commonly represented in the contemporary natural and biological sciences, *i.e.*, it constitutes "Science 101."

There prevails a virtually universal, deeply set presumption throughout our contemporary public and academic cultures that essentially all of the information traffic that stimulates and controls the multitude of interactions between these two primary domains of experience and existence propagates directly across the interface between them, with no need for alternative routing. Yet, we are confronted with an archive of irregular, irrational, yet incontrovertible empirical data that testifies, almost impishly, to our impoverished comprehension of the basic nature of these processes.

We are forced to conclude, therefore, that not only is the prevailing presumption impotent to cope with the spectrum of anomalous physical phenomena that concerns us, but it is also demonstrably inadequate to accommodate many more common aspects of human experience as well, and that these too merit re-examination from the more radical perspectives forced upon us by the anomalous behaviors.

In an effort to cope with the impasse between the subjective and objective categories of empirical experience, we have introduced two additional epistemological modules into the duplex dichotomy. On the lower left of Fig. IV-2, we have included the module Ⓤ, which encompasses those aspects of the human mind that have been labeled in various contexts and applications as "unconscious," "subconscious," "preconscious," or "non-conscious," to which have been attributed a polyglot variety of functions, including efficient storage of and access to information and past experience; autonomic control of physiological and cognitive processes; subliminal reactions to stimuli; preparation for or confirmation of those experiences which register as conscious; instinctive behavior and insight;

protection from trauma and other experiential overloads; altered states of consciousness; and various extraordinary abilities such as homing, trailing, or swarm behavior. For our present purposes, we shall subsume all such capacities and regimes, along with others implicit in this model, under a generic rubric of an undifferentiated unconscious substrate for the conscious mind, even though the contemporary psychotherapeutic community seems to hold diverse opinions regarding the most apt conceptual topology of these regimes.

Invocation of an unconscious module in and of itself would seem to benefit our model little, given the lack of empirical evidence or even plausible ideas of how this domain might share information directly with the tangible physical world any more effectively than does the conscious mind. But it is here that we take note of an increasingly common presumption of contemporary theoretical physics, namely, that there exists a somewhat comparable domain of *intangible* and abstract physical processes that underlies the tangible world, much as the unconscious mind underlies the conscious. To this we add our own radical proposition that this regime enjoys a much more intimate dialogue with the unconscious mind than does the tangible sector with the conscious, and in this form and for this function we appropriate it as yet another module of our model, (I).

Such a conceptual domain actually has been postulated in more abstract and mystical terms by natural philosophers over the full history of scientific rumination, but only recently has it attracted more orderly and analytical attention from a number of physicists. This attention has taken many parochial forms, each with its own provincial nomenclature, metaphoric imagery, and mathematical techniques, but each grasping for some sub-tangible framework for representation of an ineffable physical reservoir upon which float the tangible phenomena of observable physical events, or from which they erupt under appropriate stimulation or conditioned observation. The catchy titles of these diverse efforts are splattered throughout contemporary journals of theoretical science: "implicate order," "ontic domain," "string theories," "zero-point vacuum physics," "EPR entanglement," "quantum wholeness," "endophysics," etc., all of which struggle to capture some essence of this world of physical *potentiality.*

This shrouded, intangible domain still deals in matters of substance, energy, and information, but now presented in less distinguishable, more abstract forms that lend themselves to greater fungibility than their tangible counterparts. In some of the formalisms this loss of specificity extends into the physical metric, as well, where time and space blur and lose their functional utility. And at the deepest levels of this zone, some proponents contend, even the distinctions between mind and matter, between concept and percept, between model and data, diffuse into uncertainty,[3] and it is at this level that our own model of mind/matter intersection has some hope of completing its circuit if we can better comprehend the information transfer processes across its other interfaces.

With reference to Figure IV-3, our representation proposes that when the conscious mind expresses a strong desire, enhanced by feelings of deep resonance, that resonant intention,

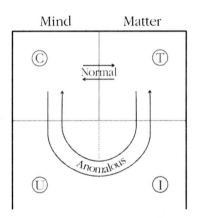

Fig. IV-3. **Indirect exchange of information between the tangible physical world and an interacting consciousness, via unconscious interactions with an intangible physical domain.**

rather than dealing directly with a tangible physical surround, stimulates some process in the unconscious mind that communicates with the pre-physical potentiality, which in turn generates a subtle biasing of the emergent tangible, albeit probabilistic, physical events. This process also may work in reverse order, as in the remote perception experience, where physical information about a target scene dissolves into an underlying intangible composition of concepts, whence it may interact with, and exert some formative influence upon, the unconscious mind of the percipient, thence to emerge into a conscious experience and subsequent re-construction of the scene.

We concede that none of the modular interfaces we propose is conceptually sharp. Rather, each entails a vague and diffuse progression of properties and processes from those of one adjacent zone to those of the other. Between Ⓒ and Ⓤ, for example, the mind progresses from fully conscious awareness to complete oblivion via various intermediate states of consciousness that have less structured or organized conceptual characteristics: autonomic control, subliminal perception, reverie, fantasy, dreaming, repressed memory, trance, hypnosis, dementia, hallucination, etc. Most of these do not switch on or off abruptly, but blur into one another in chaotic, unpredictable, and sometimes phantasmagorical mixtures. Some of them are readily accessible to conscious inspection and control, *e.g.*, breathing, heartbeat, and other forms of autonomic awareness. Other states can be broached by meditation, psychoanalysis, hypnosis, or other techniques, but many states are more deeply buried and therefore much more difficult to penetrate. And some, so far as we know, may be totally impregnable. The productive negotiation of the Ⓒ/Ⓤ, interface, therefore, is a complex and delicate task, especially if the purpose is to achieve some benign unconscious state that can establish viable communication with an amenable level of Ⓘ in the adjacent sub-physical quadrant of the duplex house.

The blurring of interface is equally evident on the material side of our modular structure, where the distinctions between tangible and intangible phenomena already are rather arbitrary, even in the classical physical and biological representations. While mechanical processes involv-

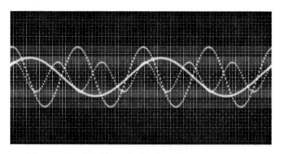

Sound propagation representation

ing the substances, configurations, motions, and interactions of discrete objects may present the appearance of unambiguous tangibility to our perceptual and conceptual senses, once we attempt to represent heat transfer and other thermodynamic effects, or the phenomena of electricity and magnetism, we inevitably are drawn into progressively more intangible abstractions of fields and waves. For example, while the wave patterns on a violin string or on the surface of the ocean may qualify as tangible in the usual sense, the propagations of sound and light waves involve less tangible properties. When we come to quantum mechanics, we have surrendered almost all claim to tangibility, certainly at the level of the wave function or state vector itself, and we are dealing with some form of potential information to be manifested only probabilistically in ⓣ. From this quantum platform, deeper progress into the intrinsically intangible formulations comprising the module ⓘ follow increasingly abstruse paths, as in the string theories wherein evanescence and uncertainty are not only tolerated but exploited, and wherein all tangible and specific coordinates ultimately disappear. (We might note that this transition also entails a growing transfer of attributed properties from common

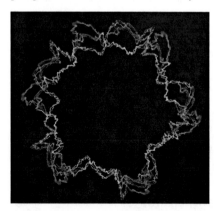

"Super-String" representation

experiential features to abstract mathematical concepts and nomenclature imposed by the theorist in the construction of his
particular model and its syntax, and in this sense mental aspects
already have permeated this material domain.)

But the most crucial interface in our model, that between Ⓤ
and Ⓘ, is the least sharply defined of all. Indeed, if the contention
of several authors regarding the indistinguishability of mental and
physical phenomena at the deepest levels of these two domains
is valid, there remains little or no interface whatsoever, simply a
pre-distinction continuum bearing only vestigial characteristics of
the divide that prevails between Ⓒ and Ⓣ. We are proposing that
it is this homogeneous, deepest layer of Ⓤ/Ⓘ that provides the
medium for anomalous passage of information from the mental
side to the material side or vice versa. Or perhaps more aptly, it
provides the gestation site for some embryonic "pre-information"
commodity that connects both tangible events and conscious experiences. Given this common origin, these palpable events and
experiences can display inherent correlations, and it is these that
comprise the apparent mind/matter anomalies that bemuse our
conscious awareness (*cf.* Figure IV-4).

Acknowledgment of the inescapably diffuse natures of the
four interfaces somewhat compromises the discrete modular

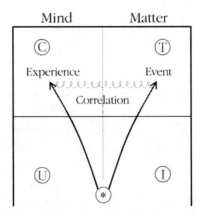

Fig. IV-4. Correlation of tangible events and experiences via shared subliminal seeds.

character of our basic model and confounds its topological representation, particularly at its central nexus. To further complicate the imagery, we should also anticipate that even these diffuse or evanescent interfaces may not be static, but rather dependent on the particulars of the participating individuals, processes, circumstances, and environments. Like the ebb and flow of the interface of the sea on the shore, what is a conscious experience on one occasion may be handled subliminally on another; what is a hard physical event in one context may be less tangibly manifested in another; and, perhaps most importantly, the extent of mind/matter distinction or merger may be keenly participant- and mood-dependent. None of this blurring or sloshing of the interfaces compromises their essential function, however. They still represent the interstices where, via each of their implicit dialogues, material, mental, and merged realities are established.

This model also raises, but does not in itself attempt to resolve, the possible role of a much vaster cosmic "Source," which could comprise a fifth, even less sharply defined module of the taxonomy, representing a composite reservoir of potentiality from which all reality derives, that permeates, influences, and informs the entire modular configuration. This is here labelled the "subliminal seed domain." If we take our cues from the tenets of most established spiritual traditions, past and present, the overarching functions of this agency, however specifically clothed, are nothing less than the creation and oversight of the material and spiritual universes which it provides for the utilization and celebration of all living creatures.

Carl Jung attempted to express a similar vision when he spoke of the multiplicity of the empirical world resting on an underlying unity, which he termed the *"Unus Mundus"* and stressed that rather than two or more fundamentally different domains existing side by side, "everything divided and different belongs to one and the same world, which is not a world of sense."[4]

While most of the characteristics of this Source domain are largely ineffable, its modes of interaction with the other conceptual modules also have some established precedents. Again drawing on the heritages of many religious, spiritual, and mystical practices, these might be catalogued as prayer, meditation, inspiration, revelation, ecstatic union, divine intervention, etc., and the pervasive powers of wisdom, morality, courage, and love, by which its teleological purpose of spiritual evolution is implemented.

A. Experimental Implications

The implications of this theoretical taxonomy for experimental design and interpretation include subtler feedback schemes that facilitate submission of conscious intention to unconscious mental processing; physical target systems that provide a richness of intangible potentialities; operators who are amenable to such implicit mechanisms; and an environmental ambience that supports the composite strategy. To illustrate the relevance of the model in the context of one of our specific laboratory experiments, consider the following scenario involving an REG device. Recall that in the absence of operator involvement (*e.g.*, calibration), the elemental binary probability of any given bit is nominally 50/50, subject to the intrinsic uncertainty of the underlying random process. When the operator first engages the machine by expressing an intention or desire for its subsequent performance, *e.g.*, "high," "low," or "baseline," this is recorded in the data manager as an *objective* experimental parameter, but it also stands henceforth as the *subjective* teleological driver of the mental side of the emergent mind/matter bifurcation. For this reverse causality to contribute to the ensuing behavior of the bonded system, however, certain attitudinal caveats have appeared to be requisite. Specifically, the operator needs to surrender conscious control of, or investment in, the desired result and submit rather to a state of detached indifference or ambivalence to the outcome, perhaps similar to that prevailing objectively in the physics of the machine, and the operator and machine need to achieve a bonded state in which their individual identities merge, thereby introducing another dimension of uncertainty into the system, much as appears to pertain in the ineffable bonds that develop between loving partners. We might note that since the very concepts of "uncertainty" and "probability," including their objective observation, inescapably entail subjective features, it may be that only within such a state of uncertainty can the elemental binary probabilities be biased and the subjective goals become objectively manifest. In other words, the mind of the operator needs to enter a "fuzzy" state where conceptual boundaries blur, categories fail, space and time evaporate, and uncertainty rules. In this merged dynamic, the initial conditions prevailing for the behavior of the material branch of the bonded system and the teleological conditions prevailing for the mental branch are playing complementary roles of equivalent importance: the former, as it were, pushing the system onward from the past, the latter drawing it forward into the future, so that its course acknowledges both its heritage and its destiny.[5]

The importance of this proliferation and manipulation of alternative states of consciousness did not escape the attention of William James. In fact, it formed the scholarly basis of one of his most notable texts:

> ...our normal waking consciousness, rational consciousness as we call it, is but one special type of consciousness, whilst all about it, parted by the flimsiest of screens, there lie potential forms of consciousness, entirely different. We may go through life without suspecting their existence; but apply the requisite stimulus, and at a touch they are there in all their completeness, definite types of mentality which probably somewhere have their field of application and adaptation. No account of the universe in its totality can be final which leaves these other forms of consciousness disregarded. How to regard them is the question — for they are so discontinuous with ordinary consciousness. Yet they may determine attitudes though they cannot furnish formulas, and open regions though they fail to give us a map. At any rate, they forbid a premature closing of our accounts toward a kind of insight to which I cannot help ascribing some metaphysical significance.[6]

Applications of M^5 to other forms of consciousness-related anomalies follow similar conceptual logic, with appropriate acknowledgment of the direction and scale of the information fluxes. With respect to the remote perception effects, for example, we are dealing with an apparently inexplicable *acquisition* of information about module Ⓣ by module Ⓒ, rather than with an apparent *insertion* of information into Ⓣ by Ⓒ. In this case we must think in terms of the physical information about the target scene being diffused through the agent's subjective experience of it into its underlying intangible composition, wherefrom it may interact with, and exert some formative influence upon, the impressionable elements of the unconscious mind of the percipient, who constructs therefrom a conscious impression and subsequent description of the scene. Considering the critical role of uncertainty in these transmutations of information, we would expect that those features of the scene that entail the least precision of specification, *i.e.*, the most generalized and impressionistic aspects, should survive this gauntlet better than features that require sharper definition.

Utilization of M^5 in the context of our FieldREG experiments presents both a confirmation and a complication. In this situation, the participants are usually unaware of the presence of the device and thus have surrendered a primary correlate of the other human/machine experiments, namely a pre-stated intention. So while it is clear that the FieldREG unit is responding to some collective mental property of the group, it is less clear what that may be. In other words, it is not obvious what here is the teleological driver. Recall that our FieldREG database divided sharply between those venues we (pre-)characterized as "resonant" or "creative," where the composite deviations of the data significantly exceeded chance expectation, and those of the more pedestrian remainder, where the results deviated significantly *less* than would be expected by chance. But such group characteristics bear no explicit relevance to the functioning of the REG electronics; the participants were not consciously attempting to produce higher or lower bit counts. Whatever their resonant intentions or desires might have been, they must have been implicit and largely un-conscious, yet they somehow influenced the degree of order in the physical environment, which in turn was reflected in the REG output probabilities.

Given the potential ubiquity of the information-sharing correl-atives just proposed, the apparent spatio-temporal independence of the phenomena requires that we add yet one more essential ingredient to the applications recipes. That is, some focus on, or specification of, the explicit task, interest, or criterion, is required to distinguish that event from all other possible correlations that could otherwise swamp the consciousness with unneeded or ir-relevant information. In the case of FieldREG, it is possible that the effects were somehow associated with the expectations, inten-tions, and/or resonant reactions of the experimenters or, wilder yet, with some collective conscious or unconscious investment of the resonant group in a purposeful outcome of the event.

Although our modular model has demonstrated some consis-tency with the bemusing bodies of empirical data accumulated in the laboratory, it can aspire to scientific credibility only if its applications lead to more replicable and significant experimen-tal results. Unfortunately, given the abstract nature of the com-ponents and configurations of the model, opportunities for such validation are quite limited and elusive to implement. Nonetheless, three categories of exploration suggest themselves. First, if we ac-cept the hypothesis that direct and explicit feedback is not nec-essarily conducive to attainment of the requisite operator states for achievement of anomalous mind/matter interactions, and that it may even be counter-productive in that it locks the process into futile attempts to penetrate the $\textcircled{C}/\textcircled{T}$ divide, it follows di-rectly that subtler forms of feedback that distract the conscious mind from the task and stimulate unconscious involvement could prove more efficacious. These might entail relaxing, numinous, or mildly hypnotic visual displays or auditory backgrounds that are not explicitly coupled to the outputs of the machines, as in our "Yantra" experiments, or a reduction of all sensory stimulation as employed, for example, in conventional "ganzfeld" experiments.[7]

Possibly most ideal would be some format that provides subliminal stimulation that is related to the operator's task or employs psychological "priming" techniques that are known to affect unconscious mental activity.[8] Whatever the form of such environmental conditioning, the operator would need to achieve and maintain a delicate balance between some teleological sense of intention or desire for a particular experimental outcome, with a surrender of conscious control or responsibility for the achievement of that goal to the unconscious mind and its deeper resources. In so doing, the operator would eschew any biofeedback-like "How am I doing?" reassurances usually provided in traditional human/machine experiments which, in this model, are hypothesized to obstruct access to the deeper unconscious intangible levels of interaction.

Another potentially effective strategy that has been suggested by operator testimony, by some abstract theoretical issues and, quite frankly, by some bald intuition, might be to establish a contradictory environment which inhibits the operator from focusing on any particular reality. For example, enigmatic images, like those depicted in paradoxical art or utilized in psychological experiments in perception, or distracting background music, could be presented to induce a bifurcated state of consciousness that some of our operators have labeled the "world between the worlds," or the "space between the bits." Or, provision of background music that is somehow inconsistent with the intended task may aid in diverting conscious intention from it. Within such an equivocal state, the unconscious mind may more readily surrender its usual conceptual reality, and merge its identity more intimately with that of the target device, in somewhat the same manner that some indigenous people merge their personalities with those of the animals and other features of their natural environments. Untainted by any preconceptions, prejudices, or consensus realities, the mind and the machine could then establish a new shared reality that would manifest as anomalous in both sectors, by our usual criteria, but would be quite "normal" in the merged state. In this bonded state, the conscious mind would not

directly query or instruct its environment; it would "dance" with it, each partner sensing and conforming to the other until a new resonant reality was formed.

Beyond the provision of more subtle feedback environments and the encouragement of operator strategies and attitudes amenable to the circuitous routes of interaction proposed by the model, some judicious selection of the random physical source and its implementation within the experimental target devices also might enhance the desired anomalous information flow. Specifically, if the properties and functions of the (T)/(I) interface are at all pertinent, it follows that physical target systems entailing complex or chaotic processes, strong non-linearities, quantum physical domains and entanglements, or any other qualities embodying high degrees of dynamical uncertainty might offer the greatest possibilities for dialogue with the corresponding unconscious mental states. Historically, these were not the targets of choice in the earliest mind/matter experiments. Zener cards, dice, and other simple mechanical devices simply did not qualify by these criteria, and it was not until the advent of electronic or radioactive REGs a few decades ago that one could systematically address the experimental efforts to truly complex sources. Here, possibly as much by blind luck as by cogent design, processes deeply rooted in quantum uncertainty underlay the tangible data streams, and more credible and replicable results were produced. Whether the growing contemporary understanding and implementations of yet more complex and indeterminate physical systems, at both the microscopic and macroscopic levels, offer options for greater resonance with the even more complex and indeterminate human mind is yet to be established, but surely should be pursued.

B. Theoretical Implications

Requisites for theoretical extension of the model include better understanding of the information dialogue between the conscious and unconscious aspects of mind; more pragmatic formulations of the relations between tangible and intangible physical processes; and, most importantly, cogent representation of the merging of mental and material dimensions into indistinguishability at their deepest levels. One possible format for mathematical treatment of the subliminal seed space that is the genesis of all mental and physical processes has been described in a companion approach labeled "M*", wherein is utilized an array of complex vectors whose components embody the pre-objective and pre-subjective aspects of their interactions.[9] Elementary algebraic arguments then predict that the degree of anomalous correlation between the emergent conscious experiences and the corresponding tangible events depends on the alignment of these interacting vectors, *i.e.*, on the correspondence of the ratios of their individual "hard" and "soft" (*e.g.* analytic and aesthetic) coordinates. This in turn suggests a subconscious alignment strategy based on strong desire, shared purpose, and meaningful resonance that is also consistent with our empirical experience.

In addition to their relevance to the results of our own experimental studies, these models also lend themselves to representation of many commonly accepted experiences, as well as to various domains of anomalous medical phenomena. For example, one might consider their application to modalities such as therapeutic touch or "energy healing," to the precognitive dimension occasionally noted in medical diagnoses,[10] or to the apparent effectiveness of individual or collective prayer in enhancing healing,[11] where in each case some emotional engagement with the patient is invoked to stimulate beneficial cellular or systemic physiological responses. They may also be pertinent to under-

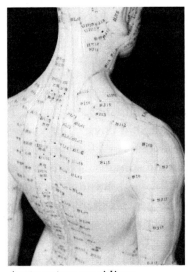

Acupuncture meridians

standing placebo responses, or the efficacy of such treatments as acupuncture or homeopathy, where stimulation of rather abstract, essentially intangible physiological information paths or processes appears to yield a variety of demonstrable clinical benefits. In all these cases, the premise would be that the tangible changes arise as material expressions of more subtle rearrangements in the relevant intangible substructures of information, themselves responsive to the unconscious imposition of the desires of the healers and the patients, utilizing the intrinsic uncertainties prevailing in both sectors.

The ultimate application of such conceptual models, of course, would be to the most pervasive of human concerns, the survival of bodily death, which we broach here with only the highest reverence and deepest trepidation, and only for those who can concede a scholarly interest in the topic. We know that as a physical corpus approaches its demise, it passes, either gradually

or abruptly, through progressively deeper stages of unconsciousness into some realm from which little tangible evidence emerges.
With the possible exceptions of neurophysiological recordings of
comatose patients or the testimonies of near-death experienc-
ers, we hold precious little evidence to define such states, but it
is not inconsistent with our model to suggest that these may be
the most propitious mental regimes from which to instigate major reconfigurations of physical, and in this case biological, reality by the means we have suggested. For example, some people
who have undergone near death experiences have reported that
these encounters have engendered substantial mental and physical changes in their subsequent lives.[12,13] In the rare cases of apparent reincarnation we may be observing an incomplete meta-
morphosis of the unconscious mind of a dying individual into a
new mortal configuration.[14] In the more common preponderance
of deaths, however, the process could be hypothesized to termi-
nate in a totally different embodiment that is essentially inaccessible to mortal observation, but nonetheless retains some aspects
of its former identity in its radically new environment, perhaps as
suggested by Dorothy Sayers in describing one possible scenario
for the Biblical resurrection.[15] The properties and parameters of
this new environment are, by definition, beyond our conscious
comprehension, but it seems unlikely that our mortal constructs
of space and time, or the distinction between mind and matter,
will remain salient when we no longer inhabit a physical environment. Whether any of this bears on various associated issues
concerning apparitions, séances, mediumship, or other forms of
purported post-mortal communication, can be pondered, but
probably not resolved.

To summarize, our purpose here has been to suggest a fresh
conceptual model of the relation between the human mind and
its material environment that can comfortably accommodate those
empirically established forms of information exchange that appear
anomalous within the prevailing theoretical paradigms. From this
proposed perspective, insights are sought that may enable more

effective and instructive experimental designs, and eventually some pragmatic applications of these inherently elusive, but potentially powerful phenomena. A major premise of this model is that further attempts at direct penetration of the interface between the universe of tangible physical events and the realm of conscious mental experience will not be productive, and even may be counter-productive, for these purposes. Rather, mind and matter must first be allowed to dissemble into less explicit and focused forms, *i.e.*, the unconscious and the intangible, to degrees where the traditional coordinates of conscious experience and tangible events lose their utility and a holistic merger of their purviews may obtain. This merged autonomous reality then may serve as a common subliminal origin from which can emerge coupled physical events and conscious experiences whose correlations exceed prevailing epistemological expectations.

References

[1] Johann Christoph Friedrich von Schiller. "Sentences of Confucius." In Sir Edward Bulwer Lytton, trans., *The Poems and Ballads of Schiller.* NY: Harper and Brothers, 1844. p. 314.

[2] Robert G. Jahn and Brenda J. Dunne. "A modular model of mind/matter manifestations (M⁵)." *Journal of Scientific Exploration, 15,* No. 3 (2001). pp. 299–329.

[3] Harald Atmanspacher and Frederick Kronz. "Relative Onticity." In *On Quanta, Mind, and Matter.* Dordrecht: Kluwer, 1999. pp. 273–294. See particularly Section 5.

[4] Carl Gustav Jung. *Mysterium Coniunctionis.* Collected Works, Vol. XIV, 2nd Ed. Princeton, NJ: Princeton University Press (Bollingen), 1970. p. 767.

[5] Antonella Vannini and Ulisse di Corpo. "A Retrocausal Model of Life." In Zachary Jones, Brenda Dunne, Elissa Hoeger, and Robert Jahn, eds., *Filters and Reflections: Perspectives on Reality*. Princeton, NJ: ICRL Press, 2009. pp. 231–244.

[6] William James. *Varieties of Religious Experience*. New York: New American Library (Mentor Books), 1958. As cited at engforum.pravda. ru/showthread.php?61576-Visions-Dreams&p=3116095

[7] Daryl J. Bem and Charles Honorton. "Does Psi exist? Replicable evidence for an anomalous process of information transfer." *Psychological Bulletin, 115*, No. 1 (1994). pp. 4–18.

[8] Bryan Kolb and Ian Q. Whishaw. *Fundamentals of Human Neuropsychology*. NY: Worth Publishers, 2003. pp. 453–457.

[9] Robert G. Jahn. "M*: Vector representation of the subliminal seed regime of M^5." *Journal of Scientific Exploration, 16*, No. 3 (2002). pp. 341–357.

[10] Larry Dossey. *The Power of Premonitions: How Knowing the Future Can Shape Our Lives*. NY: Dutton (Penguin Group), 2009.

[11] Fred Sicher, Elisabeth Targ, Dan Moore II, and Helene S. Smith. "A randomized double-blind study of the effect of distant healing in a population with advanced AIDS. Report of a small scale study." *The Western Journal of Medicine, 169*, No. 6 (1998). pp. 356–63.

[12] Raymond Moody and Elisabeth Kubler-Ross. *Life After Life: The Investigation of a Phenomenon — Survival of Bodily Death*. New York: HarperOne, 2001.

[13] Bruce Greyson. "The near-death experience scale: Construction, reliability, and validity." *Journal of Nervous and Mental Disease, 171*, No. 6 (June 1983). pp. 369–375.

[14] Ian Stevenson. *Children Who Remember Previous Lives: A Question of Reincarnation*. rev. ed. Jefferson, NC: McFarland, 2001. (345 pp.)

[15] Dorothy Leigh Sayers. Author's note to the twelfth play of *The Man Born to be King*. (Presented by the BBC December 1941–October 1942. Producer Val Gielgud.) London: Victor Gollancz Ltd, 1953. pp. 316–317.

4

SENSORS, FILTERS,
AND THE SOURCE OF REALITY*

> *Every person takes the limits of their own*
> *field of vision for the limits of the world.*
> — Arthur Schopenhauer[1]

At birth, that tiny portion of the boundless, timeless spirit of all existence that defines our deepest personal identity takes residence for one mortal span in a physical corpus we call the human body, which is given to us to experience, explore, and cultivate the sensible surround we call the world. That corpus, like the spacecraft and submersible vehicles with which we explore the physical environments of space and sea, has locomotive and manipulative capacities, and is equipped with an array of physiological sensors that can acquire specific forms of information about the environment in which it is functioning. It also possesses a neurological grid and control center, called the brain, which can be trained to interpret the signals generated by these sensors, and to activate therefrom appropriate responses. It is primarily via these channels of experience and response that we endeavor to infer, either intuitively or analytically, composite functional models of our world and of ourselves on which to base our subsequent behavior.

---

* This chapter is a condensed version of the publication in which this model was first proposed.[2] That original article prompted a number of our colleagues to attempt representations of this taxonomy from their own broad variety of professional perspectives, and the resulting essays were published in a derivative anthology entitled *Filters and Reflections: Perspectives on Reality*.[3]

The biophysical architectures and the neurological and bio-chemical processes by which these sensors and channels execute their respective functions have been broadly and deeply studied, and are sufficiently well understood that their maintenance, protection, healing, and enhancement can be practiced successfully. Yet, these physiological detectors are known to have limited ranges of sensitivity and thus ignore major segments of their corresponding spectra of stimuli. For example, human eyes perceive only the narrow band of electromagnetic radiation from 400 to 700 nanometers in wavelength, and are oblivious to the much more extensive infrared and ultraviolet domains that border it. Our ears respond only to a similarly narrow range of acoustic frequencies, beyond which lie imperceptible infrasonic and ultrasonic realms of the same form of physical oscillations. Our taste, smell, and sense of touch likewise are sensitive only to tiny portions of their potential physical or chemical stimuli. Whereas we have become technically adept at artificially extending these ranges of information access via a host of optical, electrical, and mechanical devices, our brains must then translate their outputs into extrapolations of our physiological sensitivities to effectuate their utility. We also have developed an armamentarium of equipment to amplify and refine the incoming signals for both the natural and artificially extended sensory capacities: telescopes, microscopes, hearing aids, photographic facilities, radio and television technologies, seismographs, etc. While all serve to enhance our experience of the physical world, here again our brains must execute additional steps of recognition and logic if we are to benefit from them.

More salient to our thesis here, however, is the acknowledgment that other, more subtle mechanisms for acquisition of information, such as intuition, instinct, inspiration, and various other psychical modalities, also can enhance the flux of incoming stimuli. Although commonly experienced, these channels involve less readily identifiable sensors and therefore are less susceptible to orderly reasoning and are correspondingly less respected and utilized in modern scientific and medical practice, traditional education, and contemporary social activity. In an extreme materialistic view, this imbalance extends to total dismissal of these subtler capacities, thus restricting human experience to the five primary senses and their technological extensions alone. Consequently, the inferred models of reality are limited to those substances, processes, and sources of information that are endorsed by conventional contemporary science.

Figure IV-5 displays one of several possible geometrical representations of this conceptual situation. On the left, physiological sensory channels, conditioned by various physical, psychological, and cultural filters, acquire limited information from the Source and transmit it to the consciousness for organization into palpable experience and reaction thereto. On the right, other less well-comprehended subjective channels, also conditioned by pertinent attitudinal filters, supplement the information flux with an assortment of intuitive, instinctive, and inspirational sensations that may also influence the consciousness' interpretation and response.

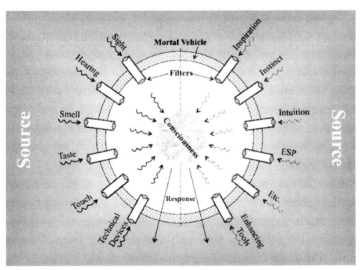

Fig. IV-5. Schematic representation of the information exchange between consciousness and its Source.

In the model offered here, we ally ourselves with the somewhat radical position that there exists a much deeper and more extensive Source of reality, which is largely insulated from direct human experience, representation, or even comprehension. It is a domain that has long been posited and contemplated in many cultures and traditions under a variety of names, none of which fully captures its sublimely elusive nature. In our M^5 model, we called it the "subliminal seed regime,"[4,5] but for our present purposes we shall henceforth refer to it simply as the "Source."

In similar spirit, we also reject the popular presumption that all modes of human information processing are completely executed within the physiological brain, and that all experiential sensations are epiphenomena of the biophysical and biochemical states thereof. Rather, we regard the brain as a neurologically localized utility that serves a much more extended "mind" or "consciousness" that far transcends the brain in its capacity, range, endurance, and subtlety of operation, and is far more sophisticated than a mere antenna for information acquisition or a silo for its storage. Indeed, we shall go so far as to contend that mind is the ultimate organizing principle that *creates* reality through its ongoing dialogue with the unstructured potentiality of the Source. In short, we subscribe to the assertion of Arthur Eddington nearly a century ago:

> Not once in the dim past, but continuously, by conscious mind is the miracle of the Creation wrought.[6]

By whatever names one might label these consciousness/ Source partners in the information dialogue, it is our conviction that the highly selective group of experiential channels based in our physiological senses allows only very limited communication between them, so that using these narrow, cloudy windows in our Source-faring capsule we can obtain only petty glimpses of the grand complexity and scope of the ultimate environment in which we are immersed, with correspondingly incomplete reflections of ourselves. Like the fabled blind men examining the elephant, our experience of the world and of ourselves is severely circumscribed by our observational inadequacies, yet it is on the basis of these superficial specifications that we presume to construct our models of reality. Even worse, these impoverished models, along with their constrained nomenclature, are then allowed to constrict further our channels of incoming information by limiting our experiences, our interpretations, and our responses, to conform to such authoritarian constructs of "reality" and to the expectations they engender. This composite dilution of experience we shall henceforth refer to as "filters," in very much the same spirit in which William Blake addressed the same issue more than two centuries ago.

> If the doors of perception were cleansed every thing would appear to man as it is, infinite. For man has closed himself up, till he sees all things through narrow chinks of his cavern.[7]

A. Two-Way Channels

That the information derived from the Source by any of our limited sensory channels is utilized by our human consciousness and its physical corpus is beyond doubt. That there is also a reverse flow of information from our mind and body to this environment is a more complex and controversial issue. Clearly, when we do something of a physical nature, *e.g.*, clap our hands, drive a car, write a book, or compose a song, our physical, cultural, or social environment is affected in some way. How much of this influence is direct physical interaction and how much is achieved by subtler subjective means is arguable, but for either form it is even less clear to what extent the deeper Source underlying the perceived environment participates in, or is modified by, this reverse information flux.

Some insight into these questions might be derived by reflecting on the role of resonance in the more conventional sensory mechanics of incoming information acquisition. Taking hearing as perhaps the simplest example, we do not merely respond to the passage of sound waves traveling down our ear ducts. Rather, some portion of those incident waves is reflected by the aural structures, thereby establishing standing wave patterns in the ducts that stimulate the drum membranes to sympathetic mechanical vibrations. These, in turn, activate the auditory neurophysiology to transmit messages to the brain for its interpretation via a corresponding pattern of standing electromagnetic waves in this segment of the channel. In other words, the ear canal, the eardrum, and the auditory portion of the brain comprise a complex acoustical/mechanical/electromagnetic resonator that is stimulated by the incoming sound signals from the environment.

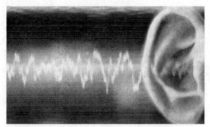 

Auditory and optical receptors

In such a model, there is an inescapable byproduct that is usually ignored as negligible, namely those portions of the outgoing reflected oscillations that reemerge from the ear aperture back into the external environment. To be sure, this entails only a tiny fraction of the energy consumed in the ear stimulation, but therein lies a mechanism for informing the environment of what the ear canal, the ear drum, and the pursuant neurophysiology, including the brain, are doing, for they are all part of one composite resonant system, the oscillations of which are available for interpretation and use by both the brain on one end, and by the environment on the other. Similar, albeit more elaborate resonant structures can be described for the eyes, mouth, nose, and tactile sensors that stimulate all manner of organic, neuronal, skeletal, and soft-tissue responses, some small portions of which are also reflected in the outgoing signals.

So far, we have not yet invoked any of the subtler forms of information acquisition mentioned earlier, but if we expand our portfolio of communication channels beyond the primary materialistic modalities to include the various subtler processes of intuition, inspiration, instinct, or the array of consciousness-related anomalous information channels, the same resonant holistic system metaphor can apply. While the specific mechanics of these incoming and outgoing information carriers and the receptors thereof may be less well understood, the role of emotional resonance in such interactions, although less tangibly defined, is impressionistically quite evident, and may well be more significant than for the physical channels. Certainly, the establishment of such resonant states between the subjective experience of a living participant and its pertinent physical and emotional environment is one of the essential functions of all spiritual practices, creative enterprises, healing strategies, and, in our context, the scientific demonstration of such anomalous resonant phenomena.

What then do such resonant channel models predicate? Clearly, the individual effect sizes are very tiny: a few phonons re-emerging from an ear or a few photons from an eye; some faint biophotonic mechanism or chemical radiation from the body; or some microscopic flux of anomalous information riding on an unspecified subjective carrier; each attempting to inform the world about the information being processed in a small portion of a brain or psyche at the other end of that particular system. Ah, but how many ears, and eyes, and mouths, and brains, and psyches, of how many living creatures equipped with such capabilities are resonating their instruments at their own particular frequencies, at any given moment? And what is the overall effect of this global concatenation of information radiations upon the world? And to what extent does that not only affect the proximate tangible environment, but also the ineffable Source from which it derives? Perhaps this imagery is akin to what William James was expressing when he referred to the mind/environment colloquy as the "aboriginal sensible muchness" out of which "attention carves out objects, which conception then names and identifies forever..."[8]

In any case, we now have in hand a conceptual framework that can employ either hard physical or soft psychical elements not only to enable ongoing two-way communication and influence between mind and matter, but also to establish a subtle network of awareness and interdependence among all of the participating minds and substances, thereby elevating the consciousness component of the information dialogue from an individual to a collective level that may merit capitalization of that term. Perhaps even more importantly, we may have identified a mechanism for individual and collective consciousnesses to access and imprint themselves on the universal Source.

At this point we might attempt to estimate quantitatively the amount of information of all physical and psychical forms currently permeating our environment that has been processed through the various resonant channels just proposed. We shall eschew that ambitious effort in favor of simply noting that any and all information that is actually extracted from the Source through the consciousness sensory systems is subject to such resonant reflections, and thereby couples these two ultimate entities to some unspecified degree. What portion of this coupling is materialistically brain-based in character, what portion is psychically based in the subtler modalities, how these two genres complement and inform one another, and how the consciousness employs its brain utility in all of this, remain major issues. But processes for consciousness to influence or condition its Source of reality do seem to exist.

We shall also pass over the multitude of possible secondary loops that connect our minds to the environment, other than to acknowledge that when we receive information through one or more of the sensory channels, mental processing may dictate a response that returns information to the environment in some other modality. When danger intrudes, I may run, resist, or perspire;

when my dog nudges me, I may speak kindly to him, pat him affectionately, or take him for a walk; when I hear great music, I may smile, sigh, or even sing along. In all cases, my reactions are returning some information to the environment, and these indirect couplings, like the more direct resonances within the individual sensory channels, greatly proliferate the grand consciousness/Source dialogue.

B. Filter Tuning

We have already noted the severe physiological restrictions imposed by our sensory equipment on the quality and quantity of information we can extract from the Source. Some of these can be improved by selective training or genetic aptitudes that favor particular modes of information acquisition, such as the enhanced sensitivities of accomplished or gifted musicians to particular tonalities; the capacities of great artists, designers, or organizers to grasp subtle nuances of shadings, designs, and patterns; or the abilities of superstar athletes to react to critical aspects of their games more vigorously and accurately than their competitors. Most of us will also concede that our prevailing subjective states of mind can color the way we perceive any events and respond to them; all of these modalities are clearly conditioned by our purposes, expectations, moods, prejudices, and attitudes, and beyond these short-term sensitivities, various longer-lived cultural and psychological filters operate as well. As an extreme example, it has been rumored that some pre-Columbian Native Americans did not see the large sailing vessels of the first European explorers to broach their shores because they had no cultural precedent for such an event or object, and no appropriate words in their vocabulary. Thus, in their context of reality, such things simply did not exist. This blindness to items that are inconsistent with expectations also has been demonstrated repeatedly under rigorous scientific conditions. For example, a number of studies in perceptual psychology have indicated that people engaged in structured activities typically do not see unexpected or bizarre events that may intrude, such as a man in a gorilla suit pounding his chest, even though these are clearly visible to uninvolved observers.[9] In other experiments, participants presented with pictures of playing cards at rapid intervals consistently misperceived incongruent cards having mismatched suits and colors.[10]

What we wish to pursue here, however, is a more proactive form of filter tuning wherein a particular physical perception, its inferred conceptualization, the patterns of conscious and unconscious response it stimulates, and the corresponding environmental reactions it induces may be altered more deliberately. This is the sort of process to which the celebrated Cherokee medicine man, Rolling Thunder, alluded when he cryptically summarized his apparent ability to manipulate external events as "There is a certain attitude you can have...."[11] We also confront it in the mystifying efficacy of the placebo, or in the demonstrable improvements in physical strength and control derivable from martial arts practices. The deliberate use of hallucinogenic substances to alter patterns of awareness and allow access to alternative realities also has extensive cultural precedents throughout recorded history.[12]

The point is that at whatever level and by whatever practice they may be invoked, such tuning techniques may condition both directions of the two-way information traffic discussed in the previous section, *i.e.*, not only may they alter the quantity and quality of the information reaching the consciousness from its Source environment, but also, via the resonant reflection processes or indirect response mechanisms just proposed, they may condition the information transmitted to the latter from the former as well. Thus, the holistic information loop of vital conversation between the two is to this extent responsive to the consciousness filters we impose, and it is to these that we must turn to seek practical benefits from our composite model. The benefits we seek, of course, are better understanding of the consciousness/Source dialogue, leading to its more effective utilization in our daily lives.

At this point the reader might reasonably anticipate some cookbook recipe for tuning the filters of consciousness to achieve

more incisive penetration into its Source environment, thereby extracting from it an expanded range of information and experience
and/or altering it in some observable way. Unfortunately, given
the fundamentally unspecifiable natures of both consciousness
and the Source, our laboratory studies have persuaded us that this
aspiration is not so neatly obtainable. Although our minds can and
do acquire information in an inherently subjective fashion, in our
Western culture they have been cultivated to conceptualize and
express their experience and activity primarily via precise "this,
not that" objective discriminations, largely neglecting the intangible subjective dimensions that can blur those distinctions. In contrast, the Source exists as a sea of ineffable, complexly intertwined
potentialities that are rooted in an intrinsic uncertainty that defies
objective specification. Thus, the same impedance mismatch that
limits our interactions with the Source also encumbers our efforts
to devise and describe means for controlling the filtering process
to enhance our access to it. This situation was eloquently summarized by Aldous Huxley in *The Doors of Perception*:

> To make biological survival possible, Mind at Large has to be
> funneled through the reducing valve of the brain and nervous
> system. What comes out at the other end is a measly trickle of
> the kind of consciousness which will help us to stay alive on
> the surface of this Particular planet. To formulate and express
> the contents of this reduced awareness, man has invented
> and endlessly elaborated those symbol-systems and implicit
> philosophies which we call languages. Every individual is at
> once the beneficiary and the victim of the linguistic tradition
> into which he has been born — the beneficiary inasmuch as
> language gives access to the accumulated records of other
> people's experience, the victim in so far as it confirms him in
> the belief that reduced awareness is the only awareness and
> as it be-devils his sense of reality, so that he is all too apt to
> take his concepts for data, his words for actual things.[13]

Hence, rather than a step-by-step recipe, we can offer only an assortment of empirical insights and derived speculations that to some may seem to be rather vague and esoteric, and to others too mechanistic and constrained. A similar disclaimer must pertain to any effort to specify the subjective strategies actually employed by the participants in our studies in generating the anomalous results that have stimulated these theoretical musings. These tactics are typically so intuitive and personal that they defy most attempts at generalization. We therefore must leave it to our readers to create their own recipes from these raw ingredients, with no assurances of effectiveness or replicability from one application to the next. Nonetheless, within these caveats, five features strike us as being salient in any productive filter-tuning strategy:

- The acceptance of the possibility of alternative representations of reality;
- The generous utilization of conceptual metaphors by which to access these alternatives;
- The achievement of an interactive resonance, in both its objective and subjective senses, with the particular targets of information;
- The tolerance of uncertainty as a *sine qua non* in any creative interaction between consciousness and its environment; and
- The replacement of conceptual duality by complementarity as the fundamental dynamic for the construction of reality.

Let us consider each of these just a bit further.

C. The Power of Perspective

The initial requisite on our list for any proactive form of consciousness filter tuning is the recognition that the reality one is experiencing is only one possible expression of the multitude of potential realities available from the Source. Only with this conviction in place is it possible to suspend or de-prioritize the particular perspective being deployed, in order to allow activation of other options. Under ordinary circumstances, however, we usually are so preoccupied with translating our experiences into objective descriptions that we fail to acknowledge at a conscious level the fundamentally subjective nature of those experiences and the accessibility of alternative representations of them. Rather, such alternatives are assigned relative probabilities via an unconscious mental algorithm that incorporates such factors as past experience, expectation, desire, or fear, along with immediate or long-term purpose or intention. As these unconscious calculations converge on the interpretation that seems most likely within the prevailing perspective, an appropriate filter is thereby imposed. Typically, it is only at this point that the attribution of experiential meaning shifts to a conscious level.

Among the many unconscious factors that contribute to such mental sorting, the prevailing values of our culture play powerful, albeit subtle, roles. The primary objectives of most socialization and educational processes are the encouragement of individual beliefs and behavior that are consistent with the values and purposes of the collective, to align our personal worldviews with the perspectives of the particular socio-cultural milieus, peer groups, or professional hierarchies in which we are immersed. Each of these dispenses its own conceptual vocabulary and priorities to bias the weighting factors in our unconscious mental calculations toward those representations of experience that are most consistent with the established beliefs and goals of that system, thereby reinforcing the coherence of its collective structure. "Thinking outside the box" tends to weaken the system's control over individual experience and action and is usually discouraged by stern social sanctions of rejection or exclusion from group membership. For humans, as well as for other social animals, such treatment is usually sufficiently painful to enforce conformity to the accepted information processing strategies, interpretations, and consequent behavior. The remarkable conformity of cadres of practitioners to the prevailing beliefs, mores, and practices of their particular branches of science are quite apt examples of such professional conformity and its subtle, and not-so-subtle, group enforcements. Eventually, these constraints become so internalized and automated that most alternative perspectives, and their associated reactions are not even recognized, let alone utilized. In extreme cases, the autonomic stress produced by the collective ethic even can engender a variety of physical and emotional pathologies, including neuroses, psychoses, or muscular armoring,[14] that further limit or distort responses to stimuli. Given such cultural obedience training, the deployment of other interpretations, *i.e.*, other consciousness filters, requires a strong act of will, plus an acceptance of the psychological consequences of deviating from the security of the collective belief system. Only then are the rules governing the creation of reality recognized as mutable rather than absolute, a realization that initially can be highly discomforting emotionally. But this is the inescapable price if one is to purchase the ability to extract from the replete potentiality of the Source an alternative physical actuality that is responsive to intentions and desires. The development of this skill is a major part of most mystical and shamanic traditions.

In such processing, the organizing mind may actually explore the application of a wide variety of conceptual filters to the boundless, undifferentiated potentiality that is the Source, whereby the emergence of several corresponding realities are possible, perhaps reminiscent of the "many worlds" concept of some physical theorists.[15] When the mind articulates an intention, it is usually associated with some sense of its meaning within a given frame of reference, and a sequence of discriminations is then initiated whereby that meaning is successively distinguished from other possible associations that that mind regards as less relevant. A familiar example of this process is the systematic solving of a crossword puzzle, where more than one word can fit a designated set of squares for which the clue is ambiguous. Even with these eliminations, however, other potential associations still remain, each of which has some residual probability of pertinence in the prevailing reference frame, thereby constituting a kind of subjective probability distribution. As each of these possibilities is considered and rejected, the variance of the distribution narrows, the selected option becomes more precise, and the mind moves from a state of relative uncertainty, to one of greater confidence.

Since most of the critical early distinctions that establish and maintain the frame of reference guiding the selection process take place at a non-conscious level, the conscious mind is rarely exposed to the less probable options. It makes its choices only among those alternatives that seem most likely, where habitual definitions of reality already have been well established. But if a different context of meaning is invoked, the distribution function of possibilities shifts to some degree, depending on how radically that alternative deviates from the conventional one. In a reference frame comprising a major change of perspective, the most likely outcome could be well out in the tails of the standard mind-set distribution; conversely, the most likely events in the latter context could be quite unlikely in the former, as in the punch line of a joke. In other words, alteration of the prevailing context of meaning to one where an ostensibly lower probability option becomes dominant is essential for effecting significant change in one's reality.

Some simple examples of such an exchange of reality would be the well-known Necker cube or face/vase illusions.

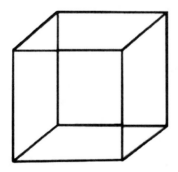 

Confronted by two equally likely perspectives, the mind finds itself in a superposition of states, shifting back and forth between the two available possibilities as it tries to determine which is the "correct" image. In such a delicately balanced situation, the slightest perceptual bias toward one interpretation or the other can shift the informational equilibrium. In more complex situations, wherein the competing interpretations have very different probabilities, however, opting for the less likely construal requires a more substantial shift of perspective, which, when achieved, engenders a correspondingly larger change in the information patterns propagating between the mind and its environment.

This may well be the controlling factor in limiting the scale of anomalous effects that characterized our laboratory-based studies. There, for successful achievement, participants had to attempt from the outset to set aside the logical "impossibility" of the assigned task and to invoke a reference frame in which the probability of the anomalous accomplishment was more likely. It is interesting to note that, in contrast to our Judeo-Christian-scientific tradition, some cultures actually encourage the exploration of alternative interpretations of experience as a strategy for developing self-awareness and deeper understanding of the relationship between the individual and the environment. Some of these traditions actually maintain that one cannot fully understand an experience until it has been considered from at least seven different points of view. In this sense, the terminology "alternative reality" might better be replaced by "multiple reality," which also befits our particular application of this concept, as evidenced by the variety of anomalous responses achieved.

D. The Magic of Metaphor

Note that throughout this chapter we have resorted to metaphor in several contexts, invoking the familiar properties of "space capsules," "distribution functions," "crossword puzzles," "Necker cubes," etc., in attempts to clarify the intended meanings of potentially unclear abstract concepts. Metaphor is a powerful technique for resolving, or utilizing, ambiguity to convey subtle nuances of meaning or to express otherwise ineffable experiences. It stimulates the associative capabilities of the mind, in contrast to its discriminatory faculties that attempt to minimize uncertainty by assigning names, categories, and functions to subjective experiences in order to reduce them to more precise objective descriptions. Discrimination, the tool of logic, presents the consciousness with sequences of "either/or" decisions that inevitably lead to a dualistic view of reality; association, the tool of creativity, raises awareness of the connections between apparently disparate interpretations and allows the consciousness to move beyond duality in representing its experience. Metaphor thus encourages the mind to pay more attention to the similarities among various interpretational options than to their differences, and thereby empowers rather than reduces the uncertainty. This is the essence of the so-called Law of Similarity that played a central role in the ancient practice of alchemy, and may lie at the heart of all forms of creativity.

Recall that in articulating the quantum mechanical model proposed in a previous chapter we had occasion to define a "consciousness atom," using a set of spherical "consciousness coordinates" that proved useful in representing various states of consciousness. For this purpose we identified "range" with the emotional intensity, "attitude" with the point of view, and "orientation" with the prevailing context, and suggested that alteration of these subjective coordinates of the mind could affect the quality of the experience by fine-tuning the resonant channels of the physical and psychical information. As another example, consider the word "green" (since cognitive scientists, philosophers of science, and theoretical physicists are fond of using color to describe qualities). Note how the consciousness coordinates adjust to the following associations: green light, green thumb, greenhouse gases, greenhorn, green about the gills, Green Man, green with envy, greenbacks, Green Bay Packers, Green Beret, Wearin' o' the Green, etc. The concept of "greenness" thus extends over a range of metaphorical implications and associated subjective meanings that goes well beyond the standard physical definition of light with a wavelength of approximately 520 nanometers. There is no single "correct" meaning, nor is there anything that precludes conscious deployment of any of the many possible meanings to color one's desired reality. This multivalent ambiguity is the essence of art, of poetry, and perhaps most ubiquitously, of humor. Virtually all jokes, cartoons, and whimsy utilize metaphor in some form, to poke fun at our foibles, fads, and follies in a friendly, unthreatening fashion that encourages fresh attention and releases the creative impulse.

Consciousness invokes the attitudinal lubricant of metaphor to facilitate adjustment of its contextual filters, thus enhancing the propagation of the incoming information from the environment. In so doing, the reflection of information back into the environment is also enhanced, and in this way the mind's intention, or any other of its subjective priorities, can be imposed on the Source, and thereby on the reality constructed from it. While we have never pursued any systematic study of the subjective strategies employed by our experimental operators, it has been evident from personal communications and casual observations that techniques of this kind are commonly deployed. These have included frank anthropomorphism, *i.e.*, attributing human qualities to the experimental devices; linguistic creation of lists of words free-associated with the prevailing intentions; or visual identification of the feedback displays with images of living processes, such as envisioning a high-intention cumulative deviation trace as a bird soaring into the air, or a low-intention trace as a fish diving into the depths. Remote perception percipients have also described their tactics in this fashion, speaking of staring at a blank screen waiting for a movie to begin, or opening a window onto a desired scene. In both classes of experiment, metaphor appears to be an effective strategy for shifting the perceived character of the task at hand from one that seems impossible to one where it is an attainable, even if unlikely, possibility.

E. The Role of Resonance

One of the most powerful and commonly employed metaphors to link the world of objective physical events with the world of subjective emotional experience is that of "resonance." Physical examples of resonant oscillatory interactions abound in all manner of mechanical, electrical, optical, and chemical systems, and characteristically entail substantial departures in behavior from those of their non-resonant counterparts. The extraordinary signal-to-noise ratios and sharp selectivities of conventional communications systems, musical instruments, and lasers of many types; the destructive oscillations of aerospace vehicles and of the Tacoma Narrows bridge; the microscale interactions that bind atoms into molecules; and the pulsations of stars and galaxies are all critically dependent on various forms of internal and external resonance.

This conceptual nomenclature is equally apt in capturing the essence of interpersonal bonds where the emergent resonant experiences, products, or performances can significantly exceed those achieved by the individual partners acting alone. It is not unreasonable, therefore, that our attempts to represent and enhance the interactions of consciousness with its environment would similarly benefit from the establishment of some form of resonance between them. The problem here is the identification of some common conceptual platform on which two apparently disparate entities may join effectively in a resonant dialogue. Our suggestion is that the requisite exchange may draw selectively on the wealth of potential information resident in the complex, chaotic uncertainty of the Source, in concert with the extensive repertoire of potential interpretations or meanings that

Ardhanārīśvara

the mind may apply. In our quantum mechanical metaphor, it was suggested that just as the binding energy of a physical molecule derives from the theoretical indistinguishability of the two participating valence electrons, the anomalous experiences that can occur among people who are "on the same wavelength," or who are engaged in resonant human/machine interactions, also may derive from the surrender of distinctions relating to individual identities in favor of a more complex "molecular" composite. We are bonded by the information we share.

To develop this proposition a bit further, we might note that any state of emotional resonance is closely associated with the perceived *meaningfulness* of the interaction. It is this visceral feeling that shifts the filters of consciousness from those of passive, objective observation to ones of proactive, subjective participation, and this participatory immersion in the interaction modifies its perceived reality. Such immersion may be enhanced by progressive elimination of the specific discriminations that distinguish self from not-self, until the consciousness approaches pure experience, a state Zen masters refer to as "samadhi." Such an experience is ineffable by definition, but those who have known it, and the traditions that have cultivated it, maintain that it approaches the ultimate reality. Some have described it as the sense of being immersed in the unmanifest potential of the universe where everything and anything is possible. Clearly, this is a very different perspective of reality than that experienced through the usual filters of perception, and its achievement may elude most of us. But many have experienced more modest forms of subjective resonance, such as being in love, where two previously independent "I's" comprise a shared "We" that can change the perception and interpretation of reality. In essence, in altering its definition of Self, consciousness attains propitious reference frames for modifying the information dialogue with its environment, and thereby enhances its ability to alter the probabilities of physical events.

The Chalice Well cover

F. The Use of Uncertainty

The prevailing ethic of virtually all scientific investigations is to strive for precision: precision of measurement, precision of analysis, sharpness of conceptualization and interpretation, maximization of the "signal-to-noise" ratio, etc. Yet, in the particular scientific context we are addressing here, this otherwise commendable zeal for precision of technique and its corresponding refinement of interpretation and prediction paradoxically can become a double-edged scholarly sword. Not only can excessive rigor in experimental design and analysis reduce rather than enhance consciousness-related anomalous effects, as revealed in our PRP studies, but there are indications that the uncertainty itself may be an essential ingredient for the generation of the anomalous phenomena. Thus we concluded that in anomalies research, as in any expression of human creativity, it is essential to establish a balance between rigor and flexibility, discipline and innovation, precision and ambiguity, if one is to navigate between the Scylla of sterility and the Charybdis of chaos.

It appears that the narrow channel that separates these complementary extremes follows an epistemological uncertainty principle similar to that which Heisenberg introduced as limiting the precision of observation of conjugate physical properties such as momentum and position, or energy and time. It is also relevant to note that the technical representation of the "zero-point field" that is postulated to permeate the universe with vast energetic potentiality,[16] ultimately derives from the imposition of this same uncertainty principle on atomic-scale harmonic oscillations. In other words, both the objective physical world and the subjective creative processes of consciousness seem to be constrained, yet enabled, by the same intrinsic ambiguity. Uncertainty inescapably characterizes the interface where the two complementary coordinate systems of mind and matter overlap to establish the interpenetration from which reality emerges. At the start of any given interaction, each of these partners resides in a state where information is still only potential, waiting for consciousness to select a frame of reference within which to address the Source, to impose appropriate subjective and objective filters, and thereby to actualize it as an experience. This done, the event can be labeled and communicated, but in so doing any alternative perspective is dismissed, and with it any information which that perspective might have conveyed. It is only in the prior phase of unresolved uncertainty, at the "margins of reality," as it were, that consciousness has provisional access not only to those realities with which it is familiar, but also to the vaster uncertainty that constitutes the ultimate Source. Progressive modification of filters with each new application further complicates the process and, given the intrinsic uncertainty that attends any chosen frame of reference, it is inevitable that all determinations of reality are inherently probabilistic. It is not surprising, therefore, that replicable results in anomalies research remain so elusive.

Notwithstanding, many experienced PEAR operators have come not only to recognize and accept this inherent uncertainty, but also to utilize it in their data-generating tactics. They speak of avoiding personal attachment to the outcome of any particular trial or run, preferring instead to "flow"[17] with the indeterminacy itself. Or as one operator expressed it, "… when it goes where I want, I flow with it. When it doesn't, I try to break the flow and give it a chance to get back in resonance with me." Successful percipients in the remote perception experiments likewise have recognized that any valid information they may acquire about their targets is convected on a background of uncertainty and possible misinformation, to which they should maintain an attitude of what has been termed "high indifference." Percipients' confidence of the accuracy of their perception was found to anti-correlate with their trial scores.

G. Conceptual Complementarity

Niels Bohr first proposed that the perplexing wave/particle paradox appearing in atomic-scale physical interactions could be rationalized only by regarding these two modes of behavior not as dualistic or contradictory, but as "complementary," in the sense that each displays a legitimate aspect of the interaction, and that both are necessary to specify it completely. Bohr himself was not averse to generalizing this physical complementarity principle into much broader philosophical and cultural dimensions: "… the nature of our consciousness," he asserted, "brings about a complementary relationship, in all domains of knowledge, between the analysis of a concept and its immediate application."[18] This extrapolation was supported by Bohr's colleagues Heisenberg and Pauli, and in their philosophical writings all three of them invoked numerous metaphors to illustrate this primary relationship.[19,20]

The magnitude of revision in conceptual and operational perspective predicated by this radical proposition should not be undervalued. Until that time, virtually all Western thought, including physics and metaphysics, had been dominated by Cartesian duality and was largely content with absolute and polar measures. Classical philosophy spoke of the dialectic tension between thesis and antithesis; theological discourse divided the world into domains of good and evil, or of spirit and matter; and this polarization of reality was reflected in the linguistic structure that underlay most common conceptualizations. In this cultural mindset, the "mind/body problem" remained at least as intractable as the "wave/particle duality." Then, from the world of hard scientific formalism, came this radical proposition that some of these sharp dichotomies should be replaced by more subtle and sophisticated complementarities, wherein specified proportions of superficially disparate properties might profitably be combined to deal with given situations. Despite its counterintuitive character, considerable theoretical elegance, as well as pragmatic benefit, in modeling both the physical world and cultural attitudes was obtained thereby.

Extension of this principle into the yet more challenging domain of consciousness mechanics can prove to be similarly beneficial. To this purpose, we have extended our metaphorical license to propose that many of the filters of consciousness that we have been addressing also may be complementary to one another in this more general sense. Grouped in appropriate pairs, such subjective/objective or receptive/assertive attitudes and perspectives entail the same orthogonal irreducibility, yet can provide the same conceptual reinforcement as the conjugate physical quantities, and can similarly serve to define the operational spaces of consciousness. It may well be that this relationship also pertains to the essence of the mind/matter interface itself, and thus to all modes of interaction between consciousness and its environment. The primary benefit of a complementary approach to the filter-tuning process is that the consciousness conjugates need not be competitive characteristics, but can be combined in appropriate proportions as befits a given situation. Participation in an experience does not preclude observation of it, nor do subjectivity and objectivity necessarily contradict one another. In fact, once the uncertainty injunction that rules out precise simultaneous specification is accepted, it may actually help define the optimal balance between any pair of consciousness conjugates.

Given our extensive musings elsewhere about the complementarity of consciousness in a variety of contexts,[21,22] we shall here add only two other examples that are particularly pertinent to the purpose of this chapter: the complementarity of intentionality and resonance, and that of consciousness and Source themselves. Intentionality and resonance both are essential in establishing the meaning of an experience, although the former is explicitly proactive and the latter intrinsically responsive. Intention imposes the filter through which the experience will be interpreted; resonance enhances the consciousness participation in the experience. Asserting an intention is essential for limiting the potential information to a given context, but it is the subjective immersion in the interaction that modifies the consciousness coordinates and thereby the meaning of the resultant information. By establishing an appropriate balance between these two conjugate states of mind, the corresponding uncertainty becomes available as a medium wherein the probabilities of possible interpretations may be altered to manifest the desired reality.

But the ultimate pair of complementary conjugates, of course, is that of consciousness itself and the ineffable Source in which it is immersed, and with which it interpenetrates to generate all manner of experience. Despite their vast disparity of conceptual character and function, it is they who comprise the universe of life, and they who are the parents of all reality.

References

[1] Arthur Schopenhauer. *Parerga and Paralipomena.* Trans. Thomas Bailey Saunders. Hayn: Berlin, 1851. Trans. Eric F. J. Payne. Oxford: Clarendon Press, 1974.

[2] Robert G. Jahn and Brenda J. Dunne. "Sensors, filters, and the Source of reality." *Journal of Scientific Exploration, 18,* No. 4 (2004). pp. 547–570.

[3] Zachary C. Jones, Brenda J. Dunne, Elissa S. Hoeger, and Robert G. Jahn, Eds. *Filters and Reflections: Perspectives on Reality.* Princeton, NJ: ICRL Press, 2009.

[4] Robert G. Jahn and Brenda J. Dunne. "A modular model of mind/matter manifestations (M^5)." *Journal of Scientific Exploration, 15,* No. 3 (2001). pp. 299–329.

[5] Robert G. Jahn. "M^*: Vector representation of the subliminal seed regime of M^5." *Journal of Scientific Exploration, 16,* No. 3 (2002). pp. 341–357.

[6] Sir Arthur Stanley Eddington. *The Nature of the Physical World.* Ann Arbor, MI: The University of Michigan Press, 1978. p. 241.

[7] William Blake. *The Marriage of Heaven and Hell.* London, ca. 1795.

[8] William James. *Some Problems of Philosophy: A Beginning of an Introduction to Philosophy.* New York: Longmans, Green, 1911. pp. 50–51.

[9] Daniel J. Simons and Christopher F. Chabris. "Gorilla in our midst: Sustained inattentional blindness for dynamic events." *Perception, 28,* No. 9 (1999). pp. 1059–1074.

[10] Jerome S. Bruner and Leo J. Postman. "On the perception of incongruity: A paradigm." *Journal of Personality, 18,* No. 2 (1949). pp. 206–223.

[11] Doug Boyd. *Rolling Thunder.* Dell Publishing, A Delta Book, 1974. p. 133.

[12] Paul Devereux. *The Long Trip: A Prehistory of Psychedelia.* New York: Penguin, 1997.

[13] Aldous Leonard Huxley. *The Doors of Perception of Heaven and Hell.* New York: Harper & Brothers, 1954. p. 23.

[14] Wilhelm Reich. *Character Analysis.* New York: Orgone Institute Press, 1949.

[15] Hugh Everett. "The Theory of the Universal Wavefunction." In Bryce Seligman DeWitt and R. Neill Graham, eds, *The Many-Worlds Interpretation of Quantum Mechanics.* Princeton Series in Physics. Princeton, NJ: Princeton University Press, 1973. pp 3–140.

[16] Bernard Haisch, Alfonso Rueda, and York Dobyns. "Inertial mass and the quantum vacuum fields." *Annalen der Physik, 10* (2001). pp. 393–414.

[17] Mihály Csíkszentmihályi. *Flow: The Psychology of Optimal Experience.* New York: Harper and Row, 1990.

[18] Niels Bohr. *Atomic Theory and the Description of Nature.* Cambridge: The University Press, 1961. p. 20.

[19] Robert G. Jahn and Brenda J. Dunne. "On the Quantum Mechanics of Consciousness, with Application to Anomalous Phenomena: Appendix B — Collected Thoughts on the Role of Consciousness in the Physical Representation of Reality." Technical Note PEAR 83005.1B, Princeton Engineering Anomalies Research, Princeton University, School of Engineering/Applied Science. December 1983, revised June 1984.

[20] Werner Heisenberg. *Physics and Beyond; Encounters and Conversations.* Tr. Arnold J. Pomerans. New York: Harper & Row, 1972.

[21] Robert G. Jahn and Brenda J. Dunne. "Science of the subjective." *Journal of Scientific Exploration, 11,* No. 2 (1997). pp. 201–224.

[22] Robert G. Jahn. "The Complementarity of Consciousness." Technical Note PEAR 91006, Princeton Engineering Anomalies Research, Princeton University, School of Engineering/Applied Science. December 1991.

5

CHANGE THE RULES!*

*A change of universal principles brings about a change of
the entire world. Speaking in this manner we no longer assume
an objective world that remains unaffected by our epistemic
activities, except when moving within the confines of a
particular point of view. We concede that our epistemic
activities may have a decisive influence even upon the most
solid piece of cosmological furniture — they may make gods
disappear and replace them by heaps of atoms in empty space.*
— Paul Feyerabend[1]

The metaphoric theoretical frameworks proposed in the three
preceding chapters are by no means trivial alterations of conventional presumptions regarding the establishment of reality. In fact,
each in its own way is at fundamental odds with the revered scientific tenets of objectivity, replicability, falsifiability, quantifiabil-
ity, causality, and determinism. Certainly, they do not entail the
usual premises and attributes drilled into young scholars aspiring
to standard scientific careers. But they are nonetheless consistent
with, indeed predicated by, the accumulated research evidence
that has repeatedly verified the existence of mind/matter phenomena that refuse to be accommodated by the traditional assumptions, and they have proven to be productive expansions of
scientific attitude for systematic study of these effects.

As it becomes increasingly apparent that objective physical
reality can be measurably affected by subjective factors that are
usually disregarded within the prevailing causal and reductionistic

---

* Our initial attempt to compose such a comprehensive assessment and
recommendation was published as an archival essay of the same title.[2]

views, it becomes correspondingly imperative that these be acknowledged and formalized if the study of consciousness-related anomalies, and indeed of consciousness itself, is to become more rigorous and productive. The most direct means for such acknowledgment would be some generalization of what heretofore have been regarded as sacred scientific principles to more comprehensive forms that explicitly include attitudinal factors, and that can be applied to these elusive but nonetheless profound and powerful phenomena. Occasionally, we have been accused of simply directing a sequence of disparate models to this task, each of which is subsequently abandoned in favor of its successor. Far from it! Each model provides a particular window on a vast and complex landscape that cannot fully be grasped from any single viewpoint. As in the parable of the blind scholars and the elephant, the systemic whole of the creature inevitably must transcend its separate parts and functions, and any individual perspective of reality is inescapably inferior to its sublime composite target.

The history of philosophy has been marked by innumerable attempts to nudge the boundaries of comprehension incrementally forward by first fracturing the chains of prevailing epistemological tyrannies in order to allow assembly of more comprehensive frameworks of conceptualization. The labels and metaphors introduced along such journeys are rarely unique, seldom sacred, and eventually supplanted, but to the extent that any conceptual model broadens a global perspective, it is probably worth contemplating, and it is in that spirit that our panoply of models has been offered. In the final Section of this book, we shall try to distill the superficially disparate features of these various efforts into at least a skeletonic conception of the full cosmic creature that recent research has exhumed from long-obscuring overlays of scientific and religious dogma, and to project its broader cultural consequences.

Before moving to that task, however, it is important to note that admission of proactive consciousness to any field of scientific endeavor, indeed to any aspect of human creativity and aesthetic satisfaction, escalates its importance to much wider constituencies. Given that the prevailing scientific paradigm, with all of its conceptual constraints and empirical shortfalls, has become the default model for virtually all contemporary scholarly, public, and personal evaluations, responses, and activities, we are obliged to contemplate the implications of our work in that vastly broader context as well.

The most fundamental and inviolate characteristic of good scientific research remains the Baconian dialogue between sound experimental evidence and astute theoretical modeling of the phenomena under study. To the dubious extent that the established scientific communities actually have familiarized themselves with the empirical results of our thirty years of anomalies research, or that of many other well-qualified laboratories before and since the PEAR era, they have tended to adopt one of the following categories of response:

1) Reject the entire body of empirical evidence as illusory.

2) Concede the anomalous effects, but dismiss their intellectual or pragmatic importance because of their small scale and elusive character.

3) Admit their existence and potential relevance, but concede the impotence of scientific methodology to deal incisively with them.

4) Attribute their irregularity and incomprehensibility to inadequate identification of additional physical factors which, once specified and controlled, would bring the phenomena into the fold of deterministic processes.

5) Relax and/or expand certain elements of scientific doctrine to encompass such phenomena in a broader scholarly paradigm.

Option #1 seems the least intellectually responsible, given the quality and quantity of the extant empirical evidence. Scraping aside the inevitable over-burden of naïve, incompetent, or fraudulent representations that have confounded the topic, the residue of solid experimental data is in fact far more extensive and incontrovertible than that undergirding many of the even more esoteric concepts of contemporary physics and biology.

Option #2 seems dangerously short-sighted in light of the numerous examples over the history of science where initially microscopic effects, once comprehended in fundamental terms, have vaulted into monumental intellectual and pragmatic importance, as in radioactivity; nuclear fission and fusion; complex and chaotic systems; nanoscience and nanotechnology; *et al.*

Option #3 needs to be considered more seriously, for it draws a line in the sand in front of contemporary science that could limit its future relevance in intellectual and public authority. In essence, it would restrict the professional purview of future science to those phenomena that submit to its increasingly rigid, mechanistic rule system, leaving assessment and understanding of all other forms of human experience to less rigorous modes of establishment, inspection, and representation.

Option #4 is reminiscent of the attitude of Albert Einstein in deferring acceptance of the probabilistic interpretations of quantum science pending a thorough search for "hidden variables" that could return the mechanics to causal terms. Unfortunately, this has yet to be empirically consummated even for most quantum-scale physical events, and certainly seems not to apply to the broader range of human experience. Yet more to the point, our established empirical data strongly indicate that at least some of such neglected correlates must be intrinsically subjective in character, for which conventional science has little capacity for specification, quantification, and mathematical manipulation. Thus, the distinction between this approach and the following option #5 essentially devolves into an attack on the subjectivity issue, *per se*, or to a reformulation of the scientific paradigm on even more comprehensive grounds.

Option #5 is clearly the most challenging, and admittedly entails two elements of substantial risk. First, there is the danger that the requisite major rule changes could be invoked illegitimately to rationalize flawed empirical data, analytical procedures, or theoretical logic, which if remedied directly would obviate the need for the more fundamental revisions. Clearly, all milder alternatives should be fully exhausted in every case before resorting to more radical resolutions. Second, if such rule changes are indeed to be installed, care must be taken not to compromise those aspects of current scientific methodology that remain valid and will continue to be essential going forward. In particular, the basic dialogue between experiment and theory must be retained, and the unavoidable loss of precision in incorporating subjective dimensions cannot be allowed to dilute the validity of the observations or the models.

All of this said, it is our considered position that some form of fundamental broadening of its strategic approach is ultimately inescapable if science is to continue to maintain its premier position in the arsenal of cultural logic and utility. If these anomalous phenomena indeed exist and can significantly impact human experience and behavior, society has the right to demand their ongoing scientific assessment and comprehension, however difficult. Beyond this, the proposed extension of the range of science would substantially alleviate several of the ideological dichotomies which its more rigid definition has fostered, *e.g.*, science vs. religion, logic vs. intuition, material vs. mental, function vs. form, etc. Two formidable obstacles obstruct this ambitious tactical path, however: the obdurate recalcitrance of the established scientific community toward this perceived softening of its "exact science" heritage; and the technical difficulties in formulating and implementing those expanded rules.

The first can be countered by noting the proliferate sequence of substantial rule changes that already have marked the course of scientific endeavor throughout its long cultural history. The recurrent pattern is for science to proceed along a relatively flat conceptual path until some imposing body of anomalous empirical evidence forces it to revise its rule system and thereby to jump rather abruptly to a new plateau of presumption, which then serves as the conceptual and strategic level for another extended era of activity until the next "paradigm shift" is forced to occur.[3] Familiar examples comprise the well-known rungs in the historical ladder of the physical sciences, frequently labeled by the names of their scholarly proponents, *e.g.*, Ptolemy to Aristotle to Galileo to Newton to Einstein to Bohr to Schrödinger, etc. From such a retrospective, we clearly have the precedents for considering another rule change and need only to argue whether the extant anomalous evidence is sufficient to merit a new shift in scientific attitude, and how this may best be achieved.

But beyond the historical argument, it should be obvious to any astute practitioner or informed observer of contemporary science that this scholarly trade is already far from a purely objective business. In choosing a topic, designing experiments, collecting and analyzing data, conceiving and formalizing models, and interpreting the ensuing dialogue between predictions and results, there is inescapable intrusion of subjective investments in the tasks at hand, acceptance of the intrinsic uncertainties lurking therein, and utilization of many forms of unconscious stimuli and processing, including motivation, intuition, instinct, and creative inspiration. While many scientists will concede, however grudgingly, the parallel presence of these factors in their daily scholarly activities, most will stoically resist attributing to them any tangible correlations with their emergent physical results. But this attitude, in the absence of any objective study of the possibilities of subjective/objective crosstalk, is itself an egregious violation of scientific objectivity. Categorical, *a priori* rejection of subjective influences in the face of the extant technical data and ubiquitous common experiences constitutes a damning indictment of the entrenched scientific establishment and its pontifical methodology.

The dereliction of science in retaining its objectivity regarding subjectivity cuts far wider than any particular research agenda addressed. Indeed, *any* scholarly discipline, in the pursuit of its own information acquisition and analyses, needs to confront the possibility that the tangible experiences in its domain may be subjectively conditioned to some degree. For this reason, we need not limit our search for precedents to the scientific milieu, but we may also examine the subjectively inspired rule changes that have characterized many other pragmatic and humanistic fields of endeavor as well. In point of fact, most of the concepts of the "exact" sciences have been metaphorically appropriated from more mundane common experience; hence their ongoing metamorphoses will likely tend to reflect the broader evolution of human thought and inspiration. Ergo, if we are searching for rule changes that can productively accommodate the subjective aspects of science, we may benefit from reflection on such influences in other realms of human affairs.

In the most common of public activities, including government and law enforcement; industry and business; education at all levels; medical practice and health care; and many other sectors, we have repeatedly installed alterations in the prevailing regulations that have reflected evolutions of public and private needs, desires, knowledge, and aspirations. In virtually every case, the issue has not so much been whether to change, but what to change, *i.e.*, what portion of the prevailing architecture of rules should remain fundamental and indispensable, and what could be beneficially modified. Current examples would include environmental protection and remediation; the ramparts erected to restrain terrorist incursions; the panoply of "equal opportunity" retrofits to societal mores; and the succession of *ad hoc* strategies to address unanticipated economic calamities.

The take-away distillation from such multidisciplinary retrospection is that quantized changes in perspectives are common, and for the most part necessary, in virtually all categories of human activity, and that scientific research in particular cannot hold itself immune from occasional overhaul when empirical evidence or philosophical maturation so predicates. Carl Jung, who faced considerable recalcitrance from his colleagues to his own unconventional ideas, conceded the difficulty of changing the scientific rules to accommodate "disturbing" questions:

This grasping of the whole is obviously the aim of science as well, but it is a goal that necessarily lies very far off because science, whenever possible, proceeds experimentally and in all cases statistically. Experiment, however, consists in asking a definite question which excludes as far as possible anything disturbing and irrelevant. It makes conditions, imposes them on Nature, and in this way forces her to give an answer to a question devised by man. She is prevented from answering out of the fullness of her possibilities since these possibilities are restricted as far as practicable. For this purpose there is created in the laboratory a situation which is artificially restricted to the question which compels Nature to give an unequivocal answer. The workings of Nature in her unrestricted wholeness are completely excluded. If we want to know what these workings are, we need a method of inquiry which imposes the fewest possible conditions, or if possible no conditions at all, and then leave Nature to answer out of her fullness.[4]

The essential tasks are to justify, program, and install such changes in constructive and expeditious fashions that minimize wastage of the prior formats, defuse resistance from the entrenched oppositions, and engender productive utilizations by the avant-garde generations of new practitioners. Beyond the representational difficulties of posing theoretical models capable of explicating and predicting physical phenomena that are correlated with subjective properties, there are further complications relating to the specification and quantification of such correlates.

But there is a still larger, even more challenging issue that must be raised in this context: namely, what is the role of consciousness in the very conception and formulation of models and theories of any genre, most especially of this one? If we are prepared to concede that consciousness can play proactive roles in the *establishment* of tangible empirical reality, are we not obliged, for philosophical consistency and symmetry if nothing else, to include and cogently represent a comparable capacity of consciousness in the theorizing process, as well? In other words, are we not now necessarily committed to *reflexive, i.e.*, self-reflecting, theories,[5] wherein the creation of models entails the same sort of dynamical intercourse of consciousness with its Source that we employ to produce the tangible physical experiences we are attempting to model? Otherwise put, have we not now moved from the creation of models of reality, to the creation of models of consciousness creating reality, to the creation of models of consciousness creating models of consciousness creating reality? Bemusing as it may seem, the inevitable consequence and high benefit of this entanglement of perspective would be to bring both our experience of reality, and our conceptual representations thereof, into synergistic indistinguishability, and thereby to avail of the most powerful technique for conditioning that reality. In a slightly different context over a century ago, Einstein succinctly opined that "It is the theory that determines what we can observe."[6] In our present context we would rephrase this to "It is our concepts that determine what we can represent." Or yet more bluntly, "Change the rules; change the reality."

Clearly, in the acceptance of these categories of rule changes, the expanded science would be surrendering some portion of its precious principles of exactness, rigid replicability, objectification, and causal determinism, to be replaced by subtler statistical criteria, consistency of correlations, metaphorical coherences, and optimized trade-offs of precision with the requisite uncertainty and randomness. That is, productive experiments and models would now strive to establish correlations of anomalous effects with both subjective and objective aspects of the protocols, not so much to trace a causal chain, but to identify those features of the procedures and environments that are propitious for spontaneous appearance of the phenomena. Astute experimenters thus would endeavor to configure their laboratories and strategies to be conducive to the emergence of effects, rather than to guarantee them, and astute theorists would cast their models in terms of enhancement of the spontaneous probabilities via the ensemble of such demonstrated correlates. The possibility that such a correlational approach might ultimately return us to causal models of the consciousness/Source dialogue is intriguing, but at this point in our understanding well beyond our grasp. Other than a few relatively superficial discussions of so-called "second-order" or "higher-level" models,[7,8] construction of fully reflexive formalisms seems to lie far ahead. But their present unattainability should not obscure our desperate need for them.

References

[1] Paul Feyerabend. *Science in a Free Society.* London: Verso, 1982. p. 70.

[2] Robert G. Jahn and Brenda J. Dunne. "Change the rules!" *Journal of Scientific Exploration, 22,* No. 2 (2008). pp. 193–213.

[3] Thomas S. Kuhn. *The Structure of Scientific Revolutions.* 2nd edition. Chicago: University of Chicago Press, 1970. p. 10.

[4] Carl Gustav Jung. *Synchronicity; an Acausal Connecting Principle.* Trans. Richard Francis Carrington Hull. 2nd ed. Princeton, NJ: Princeton University Press, 1973. 135 pp. Bollingen series no. 20.

[5] Emilios Bouratinos. *Not by Reason Alone: Prospects for a Science Toward the Limits.* ICRL Press (in preparation).

[6] Albert Einstein. In Werner Heisenberg. *Physics and Philosophy; the Revolution in Modern Science.* World Perspectives, v. 19. New York: Harper & Row, Harper Torchbooks, 1958. p. 63.

[7] George Francis Rayner Ellis. "Physics and the real world." *Physics Today,* 7 (July 2005). pp. 49–54.

[8] Harald Atmanspacher and Robert G. Jahn. "Problems of reproducibility in complex mind-matter systems." *Journal of Scientific Exploration, 17,* No. 2 (2003). pp. 243–270.

SECTION V

Consolidation and Closure

"The Puzzle"

1

INFORMATION AND INTERPRETATION

In the sense of traditional physics, information is neither matter nor energy. Rather, the concept of information brings into play the two older antipoles of matter — namely, form and consciousness.
— Carl Friedrich von Weizsäcker[1]

We come now to the two most difficult tasks this book has promised to address: the distillation of the foregoing displays of empirical evidence and theoretical propositions into some credible and pragmatic interpretation of the phenomena, and the projection of that conceptualization into a more cogent, creative, and altruistic cultural ethic than currently prevails across our diverse human societies. Both of these efforts must struggle through the entangling undergrowth of philosophical and functional dogma that has accumulated over eons of endemic human greed, self-serving rationalization, and malicious and inadvertent attentional neglect, to constrain, and often to enslave, our minds, hearts, and souls, and that has brought our species to a precipice of spiritual stagnation that cannot much longer support its survival. Our contributions here cannot be more than puny on the grand scale of such an impending catastrophe, but let us endeavor to suggest some purposeful map through this cultural jungle.

Before undertaking these last portions of our expedition, however, it might be useful to re-assemble the assortment of empirical evidence that will guide our feeble attempts to comprehend and express the implications of our foregoing work:

1) Extensive studies with a variety of random physical generators have demonstrated statistically significant correlations

between the pre-stated intentions of their human operators and the statistical output distributions of the devices.

2) The character of the effects has varied from operator to operator, but individually distinctive patterns of effects have been evident.

3) Explorations of secondary parameters, such as volitional vs. instructed assignment of intention, manual vs. automatic trial generation, modes of feedback, etc., have shown little overall correlations, but in some cases have been found to be strongly operator specific.

4) Although the scales of the observed effects have been quite small, over many repeated efforts they have compounded to highly significant deviations from chance expectations that were consistent with small shifts of the elementary binomial probabilities from their theoretical and calibration values.

5) Remote experiments, where the operators have been separated from the machines by distances of up to several thousand miles, have shown no statistical evidence of attenuation of the effects with increased distance. Similarly, no dependence of the yield on the magnitude of the time differences between device operation and operator efforts has been found over the ranges tested.

6) Significant series position patterns have been observed across diverse experiments, with tendencies for operators to produce better scores in their first series, to fall off in performance in their next few efforts, and thereafter to recover and stabilize at some intermediate levels of accomplishment.

7) Two operators working together with a shared intention have demonstrated results notably dissimilar to their individual achievements. Opposite-sex pairs, especially those sharing a personal resonance, produced more strongly significant effects than those of same-sex pairs.

8) Examination of nine distinct experimental databases have yielded consistently different patterns of performance for male and female operators, with the males tending to produce smaller

effects that correlated well with their intentions, and the females larger effects that were directionally asymmetrical and displayed larger variances in their output distributions.

9) An extensive body of FieldREG studies, deploying REGs in group situations with no pre-stated intentions as to their responses, has indicated that in venues characterized by strong group resonances the outputs display significant deviations from chance, while in more mundane environments the results compound well *below* chance expectation.

10) A database of over 650 remote perception trials has provided highly significant evidence for the anomalous acquisition of information independent of intervening distance or time.

11) Exploration of an assortment of quantitative judging techniques has demonstrated that increasing preoccupation with analytical evaluation tends to suppress the anomalous effects.

While we cannot contend that this array of experimental evidence is sufficient to specify in detail the dynamics of the underlying phenomena, it leaves little doubt that anomalous information exchange between human consciousness and randomly disposed physical devices or processes, sufficient to alter significantly their usual output behaviors, can occur in a variety of modalities. Yet, to proceed toward further understanding of these processes it is clear that more than simple patchwork repair of the prevailing scientific belief system will be needed. The ingrained presumptions regarding deterministic causation, reductionistic argument, and the sanctity of observational objectivity maintained by the entrenched majority of scientific professions and practitioners, doom any such routes to failure. Rather, while never advocating discard of any scientific methodology that has proven its beneficial utility in narrower contexts of application, we must now swallow hard and be prepared to question much of this revered logic on more profound philosophical grounds, by daring to challenge that belief system at every step, and wherever necessary, to change its rules of engagement.

In several of the preceding chapters and referenced publications we have alluded to a sequential triad of conceptual currencies routinely deployed by all of the physical, chemical, and biological sciences, *i.e.*, matter; energy; and information. We also have noted that some formal fungibility among these is well established, *e.g.*, Einstein's famous $E = mc^2$ relation concerning the transformation of mass into energy or vice versa; the information-theoretic Shannon relation, $E = kTln2$, specifying the energy required to invert one binary digit in a thermal environment; entropy definitions in classical thermodynamics; or covalent bond theories in quantum chemistry.[2] Without attempting to pursue these formats further here, it remains our considered position that of the three, *information*, in all of its ramifications, is the most facile conceptual language for depiction of the anomalous effects with which we have been engaged. The human/machine effects best lend themselves to representation as *insertions* of information into otherwise random binary strings; the remote perception results appear as *extractions* of information from a random global array of possible targets. In somewhat broader contexts, the long-standing generic phenomena of telepathy, clairvoyance, precognition, *et al.*, evidently entail information acquisition by anomalous means. Similarly, the newly labeled but long-practiced techniques of alternative healing could be regarded more parsimoniously in terms of information acquisition that helps to identify disordered physiological or psychological components (diagnosis), or information insertion that aids in balancing or re-ordering those processors (healing). The relatively recent identifications of the structures and functions of such biochemical entities as the DNA and RNA sequences, gene mappings, and brain-based neurotransmitters likewise have tended to converge on their information storage and transmission functions. Even the hallowed human propensities of prayer, trust, hope, and love essentially trade in exchanges of, and searches for, information. But more importantly, of the three conceptual currencies, information is the only one that explicitly acknowledges subjective, as well as objective, dimensions.

Indeed, if we choose to follow this information trail, it behooves us to distinguish early on between its objective and subjective forms, and then to treat these in complementary fashion. The former, the hard currency of information generation, processing, and representation of all kinds, is completely and uniquely quantifiable and, using the criteria of contemporary information science, ultimately reducible to binary digits. For example, the objective information contained in any given book could in principle be precisely quantified by digitizing each of its letters and every aspect of its syntactical structure, and compounding these in some logical schema. But the magnitude of the subjective information the book offers would still depend on the native language, the cultural heritage, and the degree of interest and meaningfulness to its readers, and thus would seem to defy precise quantification. A musical composition also can readily be digitized, but such objectification cannot adequately convey the subjective inspiration of the composer, the particular interpretation of the conductor, the nuance of articulation of the performers, or the tastes and reactions of the audience. No two performances of a piece of music are exactly alike in technique and their impressionistic expression, but their perceived quality depends on those subtle distinctions. Similarly, we might attempt to digitize the information displayed by a brilliant waterfall, a rainbow, or a great artistic masterpiece in terms of its prevailing optical frequencies and amplitudes, but in so doing we would largely fail to convey its subjective beauty or emotional impact. Nevertheless, we still may try to express in quasi-quantitative terms how much any of these sources of information affect us, *e.g.*, "This book is more interesting than that one;" "That was the worst performance I've ever heard;" or "It didn't do much for me, either way."

More precise quantification of subjective information will be a major challenge in the exploding era of information technology, particularly since it will necessitate addressing the elusive dimension of personal meaning. Some will contend that this should not even be attempted; that subjective experience must be categorically excluded from the purview of natural science. Yet science itself is subject to similar idiosyncratic influences, including the intention of the scientist, the choice of an experimental design or analytical strategy, and the interpretation of the results, all of which entail criteria, choices, and the implicit attribution of meaning that cannot be specified in purely objective terms. As William James observed:

> The spirit and principles of science are mere affairs of method; there is nothing in them that need hinder science from dealing successfully with a world in which personal forces are the starting point of new effects. The only form of thing that we directly encounter, the only experience that we concretely have is our own personal life. The only completed category of our thinking, our professors of philosophy tell us, is the category of personality, every other category being one of the abstract elements of that. And this systematic denial on science's part of personality as a condition of events, this rigorous belief that in its own essential and innermost nature our world is a strictly impersonal world, may, conceivably, as the whirligig of time goes round, prove to be the very defect that our descendants will be most surprised at in our boasted science, the omission that to their eyes will most tend to make it look perspectiveless and short.[3]

The need to include subjective information as a scientific currency is far more than an abstract philosophical issue. In a world increasingly driven by consumer marketing, media sound-bites, political posturing, and delicate interpersonal expectations, for science to deny its immense intellectual power and cultural influence to this entire domain of common human experience would not only be irresponsible, it would be self-constraining and ultimately lead to the decay of its own relevance. As we learned from our PRP research, attempts to over-objectify data risk losing valuable information by concentrating excessively on its quantitative aspects.

Imposing as this challenge of subjective representation may be, the penetration of science and technology into the jungle of meaning of information is complicated even more by the demonstrated capacity of consciousness to *alter* both the subjective and objective elements thereof. Few will quarrel with the first half of this claim. The self-evident capabilities of human consciousness to create profound experiences for itself and others to enjoy through art, music, literary composition, or even by scientific and mathematical reasoning, can hardly be disputed. The sublime moments engendered by love and empathy qualify equally well as meaningful enhancements for both donors and recipients. But the quantifiable alteration of the *objective* information content of a physical or biological system solely by the influence of an attending consciousness, which in fact has been the essence of our research, ultimately may turn out to be even more critical in optimally reconfiguring our scientific resources for service and accomplishment in the information age.

The exhortation that salient aspects of a functionally proactive subjective consciousness need to be added to the arsenal of scientific concepts and tools entails some responsibility to specify those features, and to suggest how they might productively be manipulated, integrated, and, to whatever extent possible, quantified within a comprehensive (not to mention comprehensible) "Science of the Subjective." A synoptic summary of those subjective coordinates that appear to enhance the likelihood of occurrence of consciousness-related anomalous effects emerges from the preceding text to include the following components:

Intention

Beyond simply attempting to generate some indeterminate disturbance, it seems necessary to state the intention, goal, or purpose of an interaction. That is, a specific response requires a specific question. (An exception is the FieldREG response, which has shown little correlation of its directions of deviation with any objectively specifiable parameter.[4,5])

Resonance

The union of two distinct entities into a single bonded unit, as developed in Chapter II-5, results in a "molecular" configuration with more complex characteristics and capabilities than those of its "atomic" components. Such bonding entails sharing of identity, deep trust, considerable personal investment, and genuine empathy.

Attitude

Openness and levity also appear to enable such resonance when they are propitiously tuned to those of the partner, be that a person, device, or task. It is for this reason that the PEAR investigators endeavored to maintain a warm and welcoming style in their reception of operators and encouraged them to have fun "playing" with the experiments. But the term also encompasses subtler factors such as context, meaning, purpose, and an acceptance of the unexpected.

Unconscious processing

The importance of unconscious involvement has long been evident in our experimental work and in a wider and longer range of popular lore and literature. Excessive cognitive attention, especially of an analytical, tightly focused type, may actually inhibit the state of mind that is conducive to anomalous effects. This inescapably subjects such interactions to an operational tradeoff: they need to be objectively rigorous enough to capture reliable results, but not so rigid that they suffocate the phenomena.

Gender

In many contexts, gender would be regarded as more of an objective than subjective specification. Here, however, the issue seems more a matter of masculine vs. feminine styles of information processing and priorities of accomplishment, or projective vs. receptive relationships to the device or task, rather than the usual physiological distinctions. In fact, the gender influences we have found may provide some insight into the links between consciousness and biological processes.[6]

Uncertainty

Uncertainty and probability are inseparable, and this is clearly a probabilistic game. In fact, uncertainty appears to be a fundamental essence of the Source and the raw material out of which any departures from chance are constructed. The objective/subjective tradeoff imposes an inevitable uncertainty on any constructions of consciousness, ultimately constraining them to probabilistic expressions.

Assembly of these propitious subjective qualities by no means guarantees appearance of consciousness-correlated anomalies; it does, however, enhance the likelihood of their spontaneous emergence. Conventional examples of the importance of conducive environments are self-evident: great symphonies are rarely composed in truck stops; spontaneous healing seldom occurs in overcrowded medical waiting rooms; insight and inspiration usually cannot be forced, but tend to intrude uninvited into a variety of intense personalized contexts. These anomalous events are examples of Jung's "acausal connecting principle,"[7] and the likelihood of their appearance correlates with an assortment of subjective factors. Objective conceptualization, specification, and quantification of such factors thus become unavoidably entangled. But that same uncertainty, and the opportunities for productive complementarity it presents, may well be essential to manifestations of the entire phenomenological genre and to the formulation of the new science needed to encompass it.

Clearly, the traditional array of causal, reductionistic, objective, falsifiable tenets of the prevailing scientific paradigm will require extensive softening, expansion, and circumscription to allow it to proceed profitably in this domain. Within this conviction, "Science of the Subjective" cannot be simply a cataloging mantra; it must become an endemic operational ethic that can incorporate familiar well-established phenomena along with those heretofore regarded as anomalous. What is required is the profound recognition that all of science, as most properly defined in its rigorous empirical methodology as well as in its theoretical ruminations, itself is inescapably an activity and product of the human mind, and even there is only a particularly disciplined special case of the most basic experiencing and reasoning capabilities thereof. It therefore follows that its entire expanded intellectual superstructure, both observational and contemplative, must reflect the influence of the individual and collective consciousnesses that contribute to its formation and formalization. The extent to which reality submits to the less broadly representational matrix of conventional science will depend on the particular topic, the pertinent environment, and the astuteness of the technique. Much as the Newtonian scientific perspective was ultimately forced to yield to the generalizations of relativity and quantum mechanics, so the inescapably subjective genesis of any expanded scientific logic and formalism must ultimately include evidence of that subjectivity within its representations of tangible experience.

It is our belief that here we are teetering on the threshold of another new era of science, call it "post-materialistic,"[8] "endophysical,"[9] "self-reflexive,"[10] or whatever, that will acknowledge and utilize the sublime capacities of the proactive human mind to extract far deeper aspects of physical experience from the ineffable, possibly divine Source of potential information and experience in which it is immersed. As always, that mind will continue to bring its same mental and spiritual capacities to guide it in this new formulation, but it must take great care to remain humble on its path, and not to yield to pressures of prevailing convention and ideological dogma at any stage.

References

[1] Carl Friedrich von Weizsäcker. *The Unity of Nature*. New York: Farrar, Straus, Giroux, Inc., 1980. p. 278. (Carl Hanser Verlag, Munich, 1971).

[2] Claude Elwood Shannon and Warren Weaver. *The Mathematical Theory of Communication*. Urbana, IL: University of Illinois Press, 1949.

[3] William James. "Psychical Research." In T*he Will to Believe and Other Essays in Popular Philosophy, and Human Immortality*. New York, NY: Dover Publications, 1956. p. 327.

[4] Roger D. Nelson, G. Johnston Bradish, York H. Dobyns, Brenda J. Dunne, and Robert G. Jahn. "FieldREG anomalies in group situations." *Journal of Scientific Exploration, 10*, No. 1 (1996). pp. 111–141.

[5] Roger D. Nelson, Robert G. Jahn, Brenda J. Dunne, York H. Dobyns, and G. Johnston Bradish. "FieldREG II: Consciousness field effects: Replications and explorations." *Journal of Scientific Exploration, 12*, No. 3 (1998). pp. 425–454.

[6] Brenda J. Dunne. "Gender differences in human/machine anomalies." *Journal of Scientific Exploration, 12*, No. 1 (1998). pp. 3–55.

[7] Carl Gustav Jung. *Synchronicity: An Acausal Connecting Principle*. Trans. Richard Francis Carrington Hull. 2nd ed. Princeton, NJ: Princeton University Press, 1973. 135 pp. Bollingen series No. 20.

[8] Lisa Miller, Ed. *Oxford University Press Handbook of Psychology and Spirituality*. New York, NY: Oxford University Press. (In press.)

[9] Robert G. Jahn and Brenda J. Dunne. "Endophysical Models Based on Empirical Data." In R. Buccheri, A. Elitzur, M. Saniga, eds., *Endophysics, Time, Quantum and the Subjective: Proceedings of the ZiF Interdisciplinary Research Workshop, Bielefeld, Germany, 17–22 January 2005*. Singapore: World Scientific Publishing, 2005. pp. 81–102.

[10] Emilios Bouratinos. "Primordial Wholeness: Hints of Its Non-Local and Non-Temporal Role in the Co-Evolution of Matter, Consciousness, and Civilization." In Zachary Jones, Brenda Dunne, Elissa Hoeger, and Robert Jahn, eds., *Filters and Reflections: Perspectives on Reality*. Princeton, NJ: ICRL Press, 2009. pp. 43–55.

2

CULTURAL CONSEQUENCES

*The future belongs to those who
give the next generation reason for hope.*
— Pierre Teilhard de Chardin

The rapid evolution of sophisticated computer technology over the past several decades has enabled facile acquisition and processing of new sources of complex information that have substantially altered their corresponding scientific and cultural landscapes. New categories of inquiry have arisen in virtually all dimensions of science and human affairs, and these have quickly become endemic to those domains. Even "consciousness," a term that was in considerable academic disrepute when PEAR first embarked on its journey, is now an accepted topic for serious scholarly discussion and study in several contexts. Unfortunately, many of these investigations tend to be constrained within the framework of contemporary cognitive psychology, wherein the mind is regarded as an epiphenomenon of physical brain function, albeit at some quantum level, and research strategies continue to depend on traditional scientific tools and assumptions. These, in turn, limit the range of inquiries, the types of questions that can be asked, and the answers that can be accepted. Early Newtonian reductionism, materialism, and positivism still prevail wherein the world is regarded as something outside of, and separate from, ourselves. To paraphrase an old adage, when the only tool you have is a saw, everything tends to be treated as reducible to progressively smaller components. From this perspective, it is no wonder that the study of consciousness-related anomalies remains suspect and continues to face ideological resistance by mainstream science, especially since its research results challenge the fundamental premises of those methods.

In contrast, each of the models we have proposed attempts to accommodate the array of bemusing experimental evidence by replacing conceptual duality with *complementarity*, in its broadest sense, as the primary dynamic for the construction of reality. This presumption seriously challenges many fundamental binary oppositions reflected in our contemporary science and society: *e.g.* mind vs. matter; Darwinism vs. creationism; subject vs. object; right vs. wrong; good vs. evil; etc. Even within quantum physics, wherein complementarity has long been accepted as a basic principle, the tendency is still to limit its application to specific atomic-scale physical phenomena, such as the celebrated wave-particle duality. *Uncertainty*, an inescapable adjunct of complementarity, confronts similar provincial restrictions in its applications, and like complementarity, continues to be regarded as an intrinsic characteristic of the physical world, rather than of the observer's consciousness.

Notwithstanding these limitations of focus, it has now been demonstrated that virtually all complex physical and biological systems entail components of uncertainty or randomness at their core, thereby conceding that presumptions of strict causality and determinism, along with their consequences of precise predictability or replicability, must defer to a more probabilistic posture. But if consciousness indeed has the ability to alter the probabilities of a wide range of random processes, then its proactive role in the establishment of reality must inevitably undermine the prevailing mind/matter dichotomy in favor of a more complementary integrated nature. Indeed, we may even discover that the capacity of the mind for pattern recognition and generation lies at the heart of what we perceive as the structure of the world. In his day, Arthur Eddington was quite radical in predicting that "We may look forward with undiminished enthusiasm to learning in the coming years what lies in the atomic nucleus — even though we suspect that it is hidden there by ourselves."[1] With a similar bravado, we may now look forward with undiminished enthusiasm to the evolution of a more comprehensive and interdisciplinary science that can take us beyond the study of "anomalous phenomena" to a deeper understanding of our own nature and of our relationship to the world, and to the Source of reality itself.

Some long-forgotten wag once remarked: "If we had waited until we understood combustion to build an automobile, we would all still be riding horses." Homespun as that maxim may be, it effectively conveys a universal social truth, otherwise phrased as "nothing succeeds like success." In our area of study, the practical realism is that no amount of technical evidence or erudite philosophical exhortation is likely to break down the ideological firewall between *status quo* science and the radically visionary alternative that we and others are advocating. That will occur only when the fundamental insights that have been accumulating in the basic research can be translated into demonstrably beneficial applications, such as provision of effective new therapies, attractive lines of consumer products, or useful tools and techniques for education and self-exploration. In other words, it may well be that in this field successful applications will drive the knowledge base, rather than the other way around, with profound practical, intellectual, and spiritual implications.

As such applications emerge, become accepted, and are profitably utilized by the broader popular culture, the entrenched scientific hierarchies will be challenged to respect, understand, and invest themselves in deeper comprehension of the underlying phenomena as well as their applications. The cardinal issue at this point, therefore, is whether the platform of basic knowledge and empirical accomplishment so far established in research contexts is sufficient to support these visionary enterprises despite the substantial remaining gaps in understanding. Our considered conviction is that it is, and that with adequate resources and creative proprietors, beneficial applications could indeed now be achieved. For example, the ubiquitous microelectronic REG technology that enables a major portion of the human/machine basic research is now sufficiently refined, routine, and reliable that it can readily be packaged and distributed into numerous applications arenas, supported by user-friendly instructions, service infrastructures, and readily adaptable protocols for productive and enjoyable operations. In these applications the devices could provide comprehensible streams of pertinent information; display indications of notable performances and their correlations with personal and environmental factors; and, most essentially, automatically accommodate and productively engage the subjective dimensions of the particular enterprise.

Psyleron REG

One example of innovative REG utilization that has emerged already from our basic FieldREG experiments is the Global Consciousness Project, which deploys a world-wide network of some 65 continuously operating REGs in a search for subtle correlations among themselves that may reflect the presence and activity of a collective consciousness associated with major global events.[2] It takes little imagination to visualize how similar REG devices, judiciously deployed, tended, and interpreted in other venues, *e.g.*, business planning, mass communication, public safety, education, and entertainment, etc., could suggest useful priorities for the plethora of rampant ideas and special interests prevailing in these environments. Given the well-demonstrated statistical correlations of FieldREG research data with the creative coherence generated in resonant group assemblies, it is quite likely that valuable indications of the viability of specific ideas, interpretations, or plans could be extracted over a broad range of such applications. The potential markets for an assortment of pocket-sized or table-top devices for individual self-exploration that allow users to experience, enhance, and utilize the influence of their own consciousnesses in practical applications of their personal interests, or for pure entertainment, are similarly vast.

With regard to innovative healthcare equipment and techniques, a wide range of potential applications also may wait on the horizon. As just one example, in the near future a host of tiny medical robots will very likely be deployed to clean our arteries; to inspect and service our gastrointestinal plumbing; to search for obscure growths and infections throughout our bodies; and to perform countless other diagnostic, surgical, and maintenance tasks without the need for more damaging physiological intrusions. Since applications like these will inevitably entail intentions, mind/body resonances, and the statistical uncertainties of the medical and physiological process and judgments, the involvement of consciousness-correlated anomalous effects cannot be precluded. *EXPLORE: The Journal of Science and Healing* dedicated an entire issue to the medical implications of the PEAR research, which has generated widespread interest in application possibilities within the healthcare community.[3]

Pragmatic appropriation of the concepts, data analysis techniques, and interpretive models drawn from remote perception research seems equally pertinent, and already has been explored in several venues by a number of organizations that dedicate themselves to ongoing investigation and applications of these methods. A 1996 issue of the *Journal of Scientific Exploration*[4] summarized the long-standing, but only recently declassified programs of national security surveillance of foreign activities and sites mounted by U.S. intelligence agencies, using specially trained "remote viewing" percipients and employing protocols very similar to those of our PEAR studies.

The related phenomenon of dowsing has a long and successful heritage in locating water and other natural resources, albeit anecdotally, and similar techniques have been successfully deployed in archaeological searches.[5] Some health care professionals have acknowledged such approaches in medical diagnoses and surgical procedures.[6] The use of accomplished practitioners in various law enforcement operations, most notably in searches for missing persons or property, is quietly common, if rather over-dramatized in the popular media.

We remain well aware that there are still many important scientific and strategic issues that need to be resolved if this field of study is to continue to advance, and new functional applications are to be developed. Undertaking these, however, will require much more coordinated and broadly ranging interdisciplinary programs than had been possible within the university restrictions that limited the scope of the PEAR research. Anticipating this, in the mid-1990s we began to plan how we might build on our initial accomplishments to further the development and dissemination of the insights we had gleaned, to encourage the involvement of new generations of scholars, and to begin serious consideration of pragmatic applications. An informal consortium of a few like-minded professional colleagues had earlier been assembled to discuss research issues from interdisciplinary perspectives, and this seemed a promising platform on which to build. Thus, in 1996 we incorporated this group as a not-for-profit organization called International Consciousness Research Laboratories (ICRL). In addition to several of its founding members, a number of our former PEAR interns and professional associates joined this undertaking. By the time we had closed the university laboratory in 2007, ICRL was well established and participation had grown to some 80 members from 17 countries, comprising a wide range of skills and interests relating to the interdisciplinary study of consciousness and consciousness-related anomalies.

This organization's stated goals are threefold: to extend the work of PEAR into a broader range of inquiry; to encourage a new generation of creative investigators to expand the boundaries of scientific understanding; and to strengthen the foundations of science by reclaiming its spiritual heritage. Ultimately, we seek to integrate the subjective and objective dimensions of human experience into a more comprehensive Science of the Subjective via collaborative initiatives in basic research, educational outreach, and pragmatic applications, all focused on the exploration and representation of the role of consciousness in physical and biological reality.

Beyond further investigations of anomalous human/machine interactions and various remote perception applications, ICRL activities also have addressed a broader range of issues that emerge from more diverse disciplines. For example, an investigation of the acoustical properties of ancient ceremonial sites in the United Kingdom resulted in two publications in refereed journals in the fields of acoustics[7] and of archaeology,[8] and was featured in a BBC documentary and a companion book by Paul Devereux entitled *Stone Age Soundtracks.*[9] A related collaboration explored the effects on regional brain activity[10] of the specific resonant frequencies that had been observed at these sacred sites. Other ICRL projects have addressed topics in cognitive archaeology, earth mysteries, and the properties of sacred landscapes, as well as the study of anomalous earth lights.[11-14] Studies of biophoton emission; the role of consciousness in evolution and biological processes; alternative healing modalities; the application of REG technology in psychotherapeutic situations; the integration of art and science; and human/animal communications, also have been pursued.

As part of its program of educational outreach, ICRL has established a dedicated publishing imprint, the ICRL Press, and in collaboration with StripMindMedia,[15] has produced an archival DVD/CD set, "The PEAR Proposition," and a multi-disciplinary web course on "Consciousness and the Physical World."[16] We also serve as advisors to Psyleron, Inc.,[17] a company that produces and markets REG-based technology and products.

By far the most exciting aspect of ICRL activities, however, is its interpersonal dynamics. The multi-generational composition of its membership not only promotes productive mentoring of its younger participants by those more experienced in the research

The road ahead

topics, but also introduces creative new ideas to those with more established perspectives. Periodic gatherings called "Academies" bring the group together to stimulate discussion, plan activities, and enhance the interpersonal resonance among its members.

The conclusion of the university-based PEAR era has thus led to exciting new journeys, with new possibilities, new colleagues, and new challenges, and a shared vision for the future. It is immensely gratifying to see our early efforts being advanced by a new generation of scholars, and we look forward to a productive era of new explorations. Our ambitious hope is that the integration of the subjective and objective dimensions of human experience into a new scientific perspective may help to stimulate a consequential cultural evolution wherein the academic, political, corporate, medical, and religious communities will adopt enhanced views of themselves, their purposes, their values, and their capabilities. Undergirding all of this, we envision an educational process that can break loose from its entrenchment in unproductive, stereotypical, philosophically outdated curricula and mentoring strategies into fresh formats that better prepare young minds and spirits for deeper understanding of the world and of themselves and their neighbors, and for loftier, more creative and productive lives, wherein science and spirituality ultimately will be more intimately reunited.

References

[1] Sir Arthur Stanley Eddington. *Relativity Theory of Protons and Electrons.* New York: Macmillan Co., and Cambridge: The University Press, 1936. p. 329.

[2] <http://noosphere.princeton.edu/>.

[3] *EXPLORE: The Journal of Science and Healing, 3,* No. 3 (May/June 2007).

[4] "Report on government-sponsored remote viewing programs." *Journal of Scientific Exploration, 10,* No. 1 (1996). Entire issue.

[5] Stephan A. Schwartz. *The Alexandria Project.* New York: Bantam Dell Publishing Group, 1983.

[6] Larry Dossey. *The Power of Premonitions: How Knowing the Future Can Shape Our Lives.* NY: Penguin (Dutton), 2009. p. 138.

[7] Robert G. Jahn, Paul Devereux, and Michael Ibison. "Acoustical resonances of assorted ancient structures." *Journal of the Acoustical Society of America, 99,* No. 2 (1996). pp. 649–658.

[8] Paul Devereux and Robert G. Jahn. "Preliminary investigation and cognitive considerations of the acoustical resonances of selected archaeological sites." *Antiquity, 70,* No. 268 (1996). pp. 665–666.

[9] Paul Devereux. *Stone Age Soundtracks: The Acoustic Archaeology of Ancient Sites.* London: Vega, 2001. (Re-issued, Princeton, NJ: ICRL Press, 2010.)

[10] Ian A. Cook, Sarah K. Pajot, and Andrew F. Leuchter. "Ancient architectural acoustic resonance patterns and regional brain activity." *Time & Mind: The Journal of Archaeology Consciousness and Culture, 1,* No. 1 (January 2008). pp. 95–104.

[11] Paul Devereux and Harold E. Puthoff. Marfa Lights. ICRL Technical Report #789.5. 1995.

[12] Paul Devereux. Feasibility Studies of the 'Hooker Light' Reported Phenomenon, New Jersey, and Reported Light Phenomena near Pine Bush, New York. ICRL Technical Report #93.1. 1993.

[13] Paul Devereux. Report on an Expedition to the Eastern Kimberley/ Cambridge Gulf, Australia, to Investigate Reports of Anomalous Light Phenomena in That Region. ICRL Technical Report #95.2. 1995.

[14] Bjørn G. Hauge. "Ten Years of Scientific Research of the Hessdalen Phenomena." Presentation of the Italian Committee for Project Hessdalen at an International Meeting "Le Ricerca Italiana nella Valle di Hessdalen, Norvegia," Cecina, Livorno, Italy. 2004.

[15] StripMindMedia <http://www.stripmindmedia.net/>.

[16] Consciousness and the Physical World <http://www.consciousness-studies.org/Welcome>.

[17] Psyleron, Inc. <http://www.psyleron.com/>.

3

SANDPAINTING AND SURVIVAL

I beseech you to take interest in these sacred domains
so expressively called laboratories.
Ask that there be more and that they be adorned
for these are the temples of the future, wealth and well-being.
It is here that humanity will grow, strengthen and improve.
Here, humanity will learn to read progress
and individual harmony in the works of nature ...
— Louis Pasteur[1]

The art of sandpainting has long been practiced in many cultures, and in some it serves as a deeply spiritual ceremony carried out by highly trained individuals for the purpose of healing, blessing, and facilitating transformation. The Navajo word for sandpainting, *"iikaah,"* translates as a "place where the gods come and go," and in Tibetan it is called *"dul-tson-kyil-khor,"* or "mandala of colored powders," *"mandala"* being a Sanskrit term for the world in harmony. In these traditions, the ritual usually takes place over the course of several days and involves the painstaking hand construction by one or more initiates of a complex symbolic pattern using grains of colored sand, accompanied by meditation and chant. When the ceremony has been fully completed, the painting is then dismantled and the sand swept up and dispersed to release the spirit it has embodied from its physical confines and allow it to pervade the universe. This sandpainting ritual served as an inspirational metaphor for us as the university phase of our PEAR operations drew to a close. We too had invested meticulous effort into creating a suitable scholarly program of study that had yielded a broader vision of the relationship of consciousness to its cosmic Source, and it had clearly become time to allow that vision to expand beyond its laboratory confines and academic premises.

Tibetan sandpainting

The sandpainting analogy also bears some concurrence with the structure of physical reality represented in our quantum mechanical model.[2,3] There, intangible free waves of unfulfilled probability, characterized only in terms of their frequencies and amplitudes, are attracted to, and take up residence in, physical environments represented as "potential wells," wherein they each manifest as a pattern of distinctive standing waves that reflect the qualities of both the wave and of its particular environment. These "consciousness atoms" thus synthesize the interaction of the intangible free waves and their physical environments as the source of their experiential realities. Akin to the standing waves established, say, by notes played on a musical instrument, these "consciousness eigenfunctions" define characteristic "tones" or "voices" that permeate the surrounding environment where they may contribute to a symphonic harmony produced in concert with other consciousness entities. If a particular instrument is removed from a performing ensemble, its sound is no longer physically audible, yet the echo of its voice continues to resonate in the emotions of the listeners. Similarly, the spirit of PEAR, which had pervaded its physical laboratory for many years and had drawn to it a multitude of resonant voices from beyond its physical confines, was now diffusing to extend its range even more widely.

On the occasion of PEAR's 20th anniversary in June 1999, we had organized a two-day celebration that was attended by over 150 friends of the program. Standing before this convocation of well-wishers to welcome them, we were overwhelmed by their outpouring of warmth and support and experienced our own mystical epiphany: in all the years that we had been building a rigorous scientific database, we also had been uniting this extraordinary community of people who resonated with our vision, and it was the spirit of that resonant community that may actually have been our most meaningful accomplishment. William James once observed that "mystical experiences are more than simply an amalgam of physiological, psychological, or sociological factors,"[4] and this experience certainly transcended the tangible and intellectual features of the immediate gathering. It brought to mind the profound advice of the Sage in the conclusion of Isha Schwaller de Lubicz's book *Her Bak: Egyptian Initiate:* "You are the temple which the Neter of Neters [the ultimate metaphysical force behind creation] inhabits. Awaken Him ... then let the temple fall crashing."[5] The awakening of that community spirit made us realize that a time would come when we would need to surrender our PEAR "temple," disassemble our sandpainting, and release its spirit into the world at large.

Over the past several decades we have witnessed an exponential rate of change in the physical, cultural, and personal dimensions of our lives. On the negative side, these have included planetary degradation, species extinction, natural disasters, economic instabilities, and human atrocities, to name but a few, all of which might be attributed to a growing alienation of the human spirit from its natural environment. But on the more positive side there also has been notable emergence of laudable environmental, humanitarian, and spiritual movements, as well as evidences of compassion and community in the face of tragedy. The balance between order and chaos seems to have been in exceptionally wild fluctuation, and the associated uncertainties have shaken our confidence in long-established ways of thinking. We are now faced with the choice of whether to view the present turmoil as an indication of impending apocalypse or as an emerging transformation. The comment of an old friend helped to put the situation into a more philosophical context: "I'm having a wonderful time," he told us, "just sitting back and watching the system fall apart, right on schedule."

There have, of course, been many such periods of massive change in the course of history, suggesting that our contemporary turbulence is no more than one of the recurrent series of cycles that contribute to the evolution of human and planetary development. Unlike linear representations of history, these cycles have no abrupt beginnings or endings, but they do entail periods of transition which provide precious opportunities

for personal and spiritual growth, individually and collectively. Ideally, the experience and knowledge, along with the intellectual tools and spiritual wisdom that have been gained in one cycle, may be preserved and carried forward into the next to provide for a more enlightened subsequent culture. This pertinent continuity is expressed in myths and legends like those of Atlantis, Noah's Ark, or the emergence stories of many Native American traditions, where the survivors of one cycle, having glimpsed the cosmic rainbow of the future and committed themselves to eventual fulfillment of its promise, convey the seeds of insight through turbulent periods of transition for the benefit of future generations of a transformed world.

Various hints from our PEAR studies suggest how we might effectively deal with whatever challenging times and new realities may now lie ahead. For example, our experiments indicate that alteration of only a few bits in a random distribution comprising large numbers of events can introduce some order into a physical system and shift its mean to a significant degree. We have learned that such effects can be enhanced through interpersonal resonance and that when people are drawn together by a common vision or purpose major changes in the physical environment can be achieved. The most substantial of these changes are usually accompanied by the greatest uncertainties, and are not constrained by physical space or time. And finally, the consciousness filters that we apply will determine the realities that we experience, on both the individual and collective levels, and in both the physical and spiritual dimensions.

While these insights certainly bear relevance for our collective future as a species, they also have important implications for the profound mystery that has inspired most religious traditions throughout the course of human history: the spiritual survival of bodily death. The nature of spirit or consciousness remains an enigma. The agnostically disposed regard consciousness as an emergent and cultivated characteristic of the mortal brain and therefore extinguishable at physical demise; others embrace more transcendent philosophies wherein the human spirit is the predominant quality of being, the immortal essence that relates physical existence to its cosmic origin. Certainly, the empirical evidence of non-local consciousness confounds any brain-restricted models and lends credence to a non-physical component of our being that may not be constrained to the mortal plane. Indeed, the "anomalies" we experience may be nothing less than physical manifestations of the fundamental life force expressing itself.

Nor are the implications of a non-local consciousness that can introduce order into random physical processes limited to the survival of individual personalities or endeavors. They speak as well to the bonds that connect us with one another as a living species. Like the individual cells that comprise a complex organism, we all derive from the same Source and contribute to its overall well-being. To survive periods of major cultural instability it is essential that we be able to transcend individual concerns and come together to sustain some nucleus of collective vision around which a reborn society may agglomerate. Our exhortation to "Change the Rules!", while primarily directed toward the scientific venue, extends as well to other, more comprehensive sets of rules that must be transformed if our culture as a whole is to continue to evolve. These are the rules that prevail in the unconscious and spiritual realms, the *mythos* that guides the rational *logos*, in the personal psyche and in the world at large, and that, according to Carl Jung, are driven by the archetypes of the collective unconscious.[6] Over the past few centuries our Western science has generously provided us with *logos*, the fruit of knowledge, but it has failed to offer a suitable *mythos*, the fruit of wisdom, to complement it and to infuse it with meaning that can transcend superficial appearances and counterproductive applications. It is now incumbent upon us to draw on those archetypes to create a new *mythos* that is appropriate to the transformed culture we collectively envision.

We have had a precious glimpse of this kind of intangible, yet powerful, spiritual connection in the widespread and diverse fellowship that has been drawn to PEAR over the years. Each individual has brought his or her own unique perspectives and abilities to a collective dream of the future, along with a commitment to work together to bring it into a tangible existence. And our PEAR/ICRL community is, of course, but one small component of a larger and grander kinship of visionaries that collectively, through a complementarity of intention and resonance, can compose a new, far more harmonious world symphony.

References

[1] Louis Pasteur. Statement of 1878, as quoted in Celerino Abad-Zapatero. *Crystals and Life: A Personal Journey.* La Jolla, CA: International University Line, 2002. p. 139.

[2] Robert G. Jahn and Brenda J. Dunne. "On the Quantum Mechanics of Consciousness, with Application to Anomalous Phenomena." Technical Note PEAR 83005, Princeton Engineering Anomalies Research, Princeton University, School of Engineering/Applied Science. December 1983, revised June 1984.

[3] Robert G. Jahn and Brenda J. Dunne. "On the quantum mechanics of consciousness, with application to anomalous phenomena." *Foundations of Physics, 16,* No. 8 (1986). pp. 721–772.

[4] William James. In George W. Barnard, *Exploring Unseen Worlds: William James and the Philosophy of Mysticism.* NY: SUNY Press, 1997. p. 18.

[5] Isha Schwaller de Lubicz. *Her-Bak: Egyptian Initiate.* New York: Inner Traditions, 1978. p. 248.

[6] Carl Gustav Jung. (1934–1954). *The Archetypes and The Collective Unconscious,* Collected Works, *9* (2 ed.), Princeton, NJ: Bollingen (published 1981).

EPILOGUE:
THE PEAR EXPERIENCE

Keep Ithaka always in your mind.
Arriving there is what you are destined for.
But do not hurry the journey at all.
Better if it lasts for years,
so you are old by the time you reach the island,
wealthy with all you have gained on the way,
not expecting Ithaka to make you rich.

Ithaka gave you the marvelous journey.
Without her you would not have set out.
She has nothing left to give you now.
And if you find her poor, Ithaka won't have fooled you.
Wise as you will have become, so full of experience,
you will have understood by then what these Ithakas mean.
— C. P. Cavafy[1]

Throughout this book, as in most of our earlier publications, we have emphasized the scholarly dimensions of the PEAR enterprise as the appropriate voice for communicating the experimental results and interpretations thereof of an academic research program. Yet, PEAR was much more than a conventional research enterprise; it was its people and their interactions that defined its spirit, just as the research achievements defined its substance.

The personal relationship that most strongly shaped the program, not surprisingly, was that between the two authors. Despite the vast differences in our cultural heritages, academic backgrounds, personal temperaments, and operational styles, we share a number of basic qualities. Most notably, we both tend to be intense, stubborn, opinionated, and unyielding in the face of peer pressures that oppose what we believe to be right. But we also share a common vision, along with genuine respect and compassion for each other and for all living creatures. From the interplay of these contrasts and similarities has evolved a challenging, but deeply rewarding complementarity that from the beginning has reflected a common perspective of the anomalous phenomena as much more than curious aberrations, but rather as profound indications of the nature of consciousness itself and its relationship to the grand Source of reality.

The contributions to the spirit of the PEAR lab from the other members of its staff were also substantial, particularly those of our long-term colleagues: experimental psychologist Roger Nelson, theoretical physicist York Dobyns, electrical engineer John Bradish, and our general factotum, Elissa Hoeger, each of whom devoted many years to the enterprise and collectively brought to it an array of essential skills and perspectives along with their own spiritual insights. There also were numerous transient staff members and interns who added their own proficiencies in mechanical engineering, statistics, computer programming, philosophy, comparative religion, interpersonal communications, editing and library research, office support, and administrative organization, to those of the founders' backgrounds in applied physics and the humanities, constituting a unique interdisciplinary team. Accommodation of this broad range of personalities and perspectives within the confines of a small laboratory complex, situated in a less-than-welcoming academic environment, required the cultivation of tolerance, respect, and humor, and the resulting interpersonal resonance engendered, for the most part, efficient and effective technical collaboration and warm sense of family — but also much lively debate. Birthdays, holidays, and almost any propitious occasion offered frequent opportunities for parties or picnics that enhanced this relaxed informality, and periodic retreats provided extended venues for communal discourse and interpersonal exchanges. And, of course, there was our extended family of loyal operators, a dedicated group who came to conduct experiments and frequently stayed to engage in interesting and wide-ranging conversations, contributing their own insights and suggestions.

Our relationships with our financial sponsors were similarly genial. Most of these were profoundly enlightened individuals, driven by their own personal experiences and convictions, or by intense academic fascination with the topic, to contribute to what they deemed to be a high-quality and productive research program. All of them became good friends of the program and maintained sincere personal interest in its research. Their periodic visits to the lab were always welcome occasions for stimulating discussions.

Beyond the warm and productive interactions among our staff, sponsors, and operators, over the years we also entertained a constant flow of visitors — professional colleagues who came to present ideas or to explore options for collaboration; skeptics who came to challenge our claims and to determine what it was we were "obviously doing wrong"; more casual visitors who wanted to share their personal stories with non-judgmental listeners or to obtain validation of their own anomalous experiences; students who were interested in learning about our research methods or talking about philosophical issues they may not have been comfortable discussing with their peers or faculty advisors; volunteers seeking participation in the experiments; media representatives from many countries who desired to interview us or to film the laboratory for popularized presentations; and a handful of people whose reasons for coming were not revealed explicitly, but who interrogated us with a familiar array of carefully prepared questions. Many of these visitors brought gifts of toys, stuffed animals, or art objects, and these too became integral components of the PEAR ambience.

We could write another entire book describing the multitude of interpersonal incidents that took place over the course of our odyssey, but here we shall simply share a few characteristic anecdotes — some humorous, some heartwarming, some bewildering — that may help to convey a sense of these intangible, yet meaningful, dimensions of the PEAR experience.

Synchronistic events, such as receiving phone calls from people just as we were talking about them, or technical malfunctions occurring during periods of high stress or intense urgency, were sufficiently common that we tended to take them for granted. Some incidents, however, entailed more unusual anomalous components that were as intriguing in their own way as the formal experimental evidence. As one example, in the early years of the program a somewhat unsavory individual, whose agenda was unclear but whose presence in the laboratory made some members of the staff uneasy, visited for several days. During this period there were a number of research activities in progress, including a series of PRP trials with a colleague who was engaged at the time in a single-handed sailing trip across the Atlantic Ocean. In this percipient's PRP transcript for one of the days in which our strange visitor was present, she alluded to her sense of a sinister "voyeur" who felt "threatening," and reported that "overall I felt a strong connection of apprehension for Brenda — that she was being watched to no good purpose." At this time we were also in the process of running a continuous calibration of our Fabry-Perot interferometer device to determine its sensitivity to environmental temperature fluctuations. Curiously, whenever this stranger entered the laboratory the chart recorder that was monitoring room temperature began to produce erratic traces. After several such episodes, he became evidently disconcerted and suspiciously demanded to know what it was we were actually measuring. Our truthful response, offered with a smile and casual shrug, was "temperature cycles," to which his eyes grew wide and his face grew pale, and he left hurriedly, slamming the door behind him, never to return. The chart recorder thereafter returned to its normal operation and we never again saw a repetition of its odd behavior.

Visits and letters from "professional skeptics" were fairly common. In one case a skeptic who challenged our claims of nonlocal REG effects was encouraged to try the experiment for himself by participating as a remote operator. A series was arranged wherein he was to carry out sessions of 1000 trials per intention on five successive days, and at the end of each session he was to fax to us the order of his three intentions and we would then fax to him the day's results. Over the first four days his results consistently displayed exceptionally strong trends *opposite* to his stated intention, and although we offered no commentary on these it was as clear to him as it was to us that they were compounding to a result well beyond chance. On the final day of the experiment, however, he produced an extremely significant positive-going effect of about four standard deviations in magnitude — just sufficient to bring his final total yield to a terminal probability of precisely .50. We heard nothing further from him, but imagined him heaving a strong sigh of relief.

On another occasion, two well-known skeptics agreed to participate as percipients in an informal PRP trial, and each was provided with a portable tape recorder into which he was to describe his impressions of the target. One produced a strikingly accurate representation of the target site — a backyard patio and swimming pool — but part way through his rendition he broke into an alarmed description of a bee that had been flying around his head and had somehow gotten under his glasses. The other percipient did not produce any information relevant to the target scene, but instead described an image of a man dressed in a bumblebee costume!

Another instance involved a visit to the laboratory by a prominent member of the skeptical community who had been invited by the university administration to present a distinguished endowed lecture. Despite his thorough inspection, he was unable to find any evidence of naïve or fraudulent activity in our procedures or equipment. Yet during his public lecture that evening, the members of the PEAR staff, who were seated in the front row of the audience, were amused by an ongoing sequence of malfunctioning microphones, slide projectors, and lights that repeatedly plagued his presentation.

A frequent argument raised by some critics of the PEAR program was that our work could not be taken seriously until it had been published in refereed mainstream science journals. Given that the majority of such journals categorically refused to consider papers on this topic, this issue became something of a "Catch-22," and our interactions with journal editors provided yet another source of interesting anecdotes, frustrating at the time, but amusing in retrospect. For example, shortly after the program was established we were invited by the editor of a prominent engineering journal to submit a review article. That article, entitled "The Persistent Paradox of Psychic Phenomena,"[2] generated an enormous response from readers, as indicated by more than a thousand requests for reprints and its translation into several foreign languages. Yet several years later, when we contacted this same editor to ask if he would be interested in considering a follow-on article, we were told that the readers of his journal had no interest in the subject. At another point we submitted an article describing our quantum mechanical model to a respected theoretical physics journal. In the usual course of the peer review process, a paper typically is evaluated by at most two or three reviewers, but this one was sent to *seventeen* readers and elicited a broad range of both positive and negative responses. After deliberating for more than a year, the editor finally decided simply to make his decision on the basis of a majority vote, and the paper was ultimately published,[3] but as in the previous example, we were later informed that no subsequent article would be considered. And then there was the editor of a prominent science magazine who informed us that he would consider publishing an article about our work only when we were able to transmit the manuscript to him telepathically!

On more than one occasion, editors of other distinguished academic journals solicited articles that subsequently were summarily embargoed by their editorial boards with little or no scholarly justification. A comprehensive technical article summarizing the results of twelve years of our human/machine experiments was sequentially submitted to three different scholarly journals, each of which refused to consider it on transparently superficial grounds. Along with most of our other archival articles, this was eventually published in the *Journal of Scientific Exploration*,[4] one of the very few peer-reviewed professional journals where submissions are evaluated solely on the basis of their quality, regardless of their subject matter. Two well-known science journals made it clear that they would not be interested in any submissions we might make, yet both promptly printed announcements of our laboratory closing in 2007, one of which bore the headline "The Lab That Asked the Wrong Questions".

Our numerous presentations to professional societies, academic and government seminars, and public meetings were usually courteously received, and the subsequent question and answer periods typically consisted of polite, but often mildly skeptical reactions. After the formal sessions adjourned, however, we were almost invariably approached privately by individuals who would glance over their shoulders, lean in closer, and in lowered voices inform us that they were really very interested in what we were doing and that they actually had experienced some anomalous events of their own which they were not comfortable confiding to others. Many of these individuals subsequently maintained correspondences with us, visited the lab, or even made financial contributions to support the program.

On the whole, our ongoing interactions with the university community tended to be rather testy, and from the outset our enterprise was regarded with suspicion. For example, as one of the conditions for its approval, it was subjected to the oversight of an *ad hoc* committee, comprising all of the senior academic officers of the institution, including the President, Provost, Dean of the Faculty, Dean of the College, Dean of the Graduate School, and Chairman of the University Research Board — an extraordinary assemblage for such a modest-scale research program funded solely by philanthropic gifts!

This prevailing discomfort occasionally produced some rather droll indications of the nature of the wide-spread apprehension that our presence elicited. For instance, when the laboratory was first established, in what had formerly been a storage area with no identifying room number, we had temporarily posted an image of a Greek psi (Ψ) on its door, intended to suggest the multiple symbolism of a quantum mechanical wave function, the field of psychology, and that of psychic phenomena. Over the next several weeks, we received at least three inquires, accompanied by troubled expressions, asking why we had a devil's pitchfork on our door.

We were similarly amused by the response of the university's Human Subjects Committee to our request for approval of our experimental equipment and protocols. Authorization was eventually granted, but with the caveat that "the researchers be sensitive to the possibility of inducing an unwarranted self-perception of metaphysical capabilities on the part of the subjects, and employ methods appropriate to minimizing the probability of this occurring." This ridiculous injunction actually proved beneficial. Whenever we would carefully inform our operators that we had been instructed by the university to warn them against interpreting the experiments as producing "an unwarranted self-perception of metaphysical capabilities," they would inevitably laugh, thus setting the playful tone we regarded as propitious for successful performance. (This imposition was also one of our reasons for referring to our experimental participants as "operators," rather than "subjects.")

Many of our colleagues openly disapproved of what we were doing, or what they thought we were doing, since virtually none of these visited the lab or made any evident effort to inform themselves about the actual substance of our research. Nevertheless, these individuals tended to be quite outspoken in their censure, including rationalizing their position with the excuse that we were compromising the university's public image by engaging in a program of "pseudo-science." On one occasion we received a visit from a physics graduate student who had been so taken aback by his advisor's uncharacteristically intense emotional rejection of our work that he decided to investigate it for himself by becoming an operator.

On the other hand, there also were some colleagues who, at least privately, expressed sincere interest and on occasion would stop by to talk or to even to participate in the experiments (although they frequently requested that we not reveal their involvement to their co-workers or supervisors). A few members of our faculty were even more openly supportive, and some were genuinely helpful, especially in the early days of the program, providing us with encouragement and valuable technical advice.

In 1986, with financial support from the McDonnell Foundation, we were able to establish an interdisciplinary academic program known as the Human Information Processing (HIP) Group that brought together faculty members and research staff from four university laboratories: the Cognitive Motivation Laboratory, the Cognitive Science Laboratory, the Robotics and Expert Systems Laboratory, and PEAR; for the purpose of studying "all aspects of human-machine interaction, encompassing such diverse research topics as the optimal design of user-computer interfaces, the limits of machine and human intelligence, the design of expert robotic systems, the cognitive processes involved in motivation and reasoning, anomalies arising in human-machine interaction, and the nature of consciousness itself." Although the research collaborations that emerged from this program were limited, in 1987 the group initiated a popular team-taught undergraduate course that was offered annually for ten years and provided the sole curricular opportunity for us to share our work and ideas with typically 50 students per year drawn from various academic departments. This course included five hours of lectures explicitly describing the PEAR program, a visit to the PEAR laboratory during which the students had an opportunity to see and try the experiments, and a requirement to conduct an independent research project in connection with one of the participating laboratories. Many of the projects selected by the students were based on PEAR-related topics and several of those pursued were of exceptionally high quality. In addition to the HIP student projects, a number of other Princeton undergraduates were able to obtain approval to conduct independent research projects with us, several of which were of sufficient distinction to stimulate continuing formal laboratory studies, and in some cases, to merit eventual publication.

Our educational outreach also extended well beyond Princeton to involve numerous students from other institutions, ranging from elementary school children to doctoral and post-doctoral scholars. For eight years we hosted annual visits of several classes of "gifted and talented" 4th-grade students from a nearby elementary school district as part of an introductory science class in which they were required to design and carry out original science experiments of their own design. Since

Probability lesson

we were apparently the only academic science laboratory in the area willing to commit the time and effort these visits required, these student projects all addressed experiments in consciousness-related anomalies. They would come in groups of 10 to 15 at a time and spend a morning for a lecture on scientific method, probability, and experimental design; trying our experiments; and best of all, having a chance to present and describe their own projects. Over the years, as their teachers became more familiar with the nomenclature, concepts, and principles underlying our work, their students' projects became more sophisticated, creative, and well-designed, and such visits became a highlight of each year, both for them and for us. When the laboratory closed several years later, we received a moving email from one of them, by then a graduate student, wishing us well in our future endeavors and observing how meaningful her visit back in fourth grade had been.

"Murphy, go right!"

Our most productive and rewarding student involvements, however, came in the form of a long line of visitors, ranging from high-school to post-graduate level, who spent anywhere from a few days to several months as guests and active participants in our program. Many of these have since continued to maintain regular contact with us and with each other as they have moved on to their own wide-ranging careers. The members of this treasured community, which came to be known as the "PEARtree," have remained valued colleagues in our extended ICRL family and comprise a second generation of scholars who are carrying forward the PEAR heritage in many innovative directions. When the time came for us to close the university laboratory, several of them returned to participate in that event, and each took home one of our stuffed animals as a keepsake. At that time, they presented us with a beautiful book of pictures and profound personal tributes, including a moving dedication that read:

> For this family, PEAR was more than just a laboratory. It was a cherished community where we could express ourselves among friends and like-minded peers. For some, it was a place of inspiration where we found the inner strength to pursue our dreams. For others, it was a haven to escape from the expectations of a judgmental society. In truth, the PEAR laboratory probably represented something a little bit different to every visitor who stepped through the door. Yet the spirit of

PEAR shined clearly for each of us. PEAR — the laboratory, the people — has been a place of Love and Spirit, Laughter and Tears. In your company we found Truth and Honesty, Kindness and Motivation. It was a second home, a chosen family whose influence we feel daily. It is Magic.

Bob and Brenda, this album is our small gift, meant to tell you of our tremendous appreciation for the work you have done in the world and in our lives. We hope that looking through this book you will smile, cry, and remember all the wonderful times we had together. But most of all, we hope that you will see how clear it is to us all that the closing of the PEAR lab is also a beginning. Out of the planter and into the soil, PEAR will continue to thrive, growing new leaves with each season, strengthening its roots, bearing fruit to feed the hungry, siring other Trees, and sprouting new branches that will reach ever closer to Heaven.

It doesn't get any better than that!

References

[1] Constantine Petrou Cavafy. "Ithaka." In George Savidis, ed., trans. Edmund Keeley and Philip Sherrard, *Collected Poems*. Revised Edition. Princeton, NJ: Princeton University Press, 1992.

[2] Robert G. Jahn. "The persistent paradox of psychic phenomena: An engineering perspective." *Proceedings IEEE*, *70*, No. 2 (1982). pp. 136–170.

[3] Robert G. Jahn and Brenda J. Dunne. "On the quantum mechanics of consciousness, with application to anomalous phenomena." *Foundations of Physics*, *16*, No. 8 (1986). pp. 721–772.

[4] Robert G. Jahn, Brenda J. Dunne, Roger D. Nelson, York H. Dobyns, and G. Johnston Bradish. "Correlations of random binary sequences with pre-stated operator intention: A review of a 12-year program." *Journal of Scientific Exploration*, *11*, No. 3 (1997). pp. 345–367.

ILLUSTRATION CREDITS

Front cover, Compliments of Odd Erik Garcia

PROLOGUE:
The PEAR Experience

P. x, Robert Jahn and Brenda Dunne

SECTION I
Venues, Vistas, and Vectors

P. 1, Dreamstime <http://www.dreamstime.com>; *p. 4,* Dreamstime; *p. 8,* Dreamstime; *p. 17,* public domain; *p. 18,* public domain; *p. 24,* Courtesy of Christian Skeel and Morten Skriver; *p. 28,* public domain; *p. 29,* http://www.fyndo.com/teniers-alchemist.jpg; *p. 35,* Courtesy of Nisha Checka <nisha.checka@gmail.com>

SECTION II
Human/Machine Connections: Thinking Inside the Box?

P. 41, Robert Jahn and Brenda Dunne; *p. 54,* Robert Jahn and Brenda Dunne; *p. 59,* Robert Jahn and Brenda Dunne; *p. 59,* Robert Jahn and Brenda Dunne; *p. 81,* Robert Jahn and Brenda Dunne; *p. 83* left, Rose Petals and Mussel Shells, Driftwood Beach, Great Spruce Head Island, ME, July 1971. Photograph by Eliot Porter in *Nature's Chaos,* Ed. Janet Russek, NY: Viking Penguin, 1990, p. 101; *p. 83* right, Bear Screen, Artist of the Stikine division of Tlinget, Denver Art Museum, Denver CO. From *An Encyclopedia of Archetypal Symbolism,* Ed. Beverly Moon, Boston and London: Shambala, 1991, p. 94; *p. 86,* Robert Jahn and Brenda Dunne; *p. 86,* Robert Jahn and Brenda Dunne; *p. 106,* Robert Jahn and Brenda Dunne; *p. 109,* Colin Faulkingham; *p. 122,* Galton Laboratory, University College, London; *p. 123,* Robert Jahn and Brenda Dunne; *p. 125,* Robert Jahn and Brenda Dunne; *p. 131,* Robert Jahn and Brenda Dunne; *p. 133,* Robert Jahn and Brenda Dunne; *p. 136,* Robert Jahn and Brenda Dunne; *p. 141* both, Robert Jahn and Brenda Dunne; *p. 154* both, Robert Jahn and Brenda Dunne; *p. 156,* Robert Jahn and Brenda Dunne; *p. 159,* Robert Jahn and Brenda Dunne

SECTION III
Remote Perception: Information and Uncertainty

P. 195, courtesy of A. Andrew Gonzalez <http://www.sublimatrix. com>; *p. 206,* Robert Jahn and Brenda Dunne; *p. 226,* Robert Jahn and Brenda Dunne; *p. 230,* Robert Jahn and Brenda Dunne; *p. 234* both, Robert Jahn and Brenda Dunne

SECTION IV
Thinking Outside the Box

P. 257, Dreamstime, © Mike301; *p. 258,* Courtesy of Zachary Jones, 2006. <www.KeystoneFlow.com>; *p. 268,* Dreamstime; QUARTZ CRYSTALS by Tektite; *p. 272,* Google images; *p. 276,* Dreamstime; *p. 276,* Dreamstime; *p. 283,* Dreamstime; © Billyfoto | Dreamstime.com; *p. 286,* Dreamstime; ACUPUNCTURE CONCEPT by Corachaos; *p. 295,* Dreamstime; Garygrc02; *p. 304* both, public domain; *p. 305,* public domain; *p. 309,* photographed by Adam Curry; *p. 310,* Robert Jahn and Brenda Dunne

SECTION V
Consolidation and Closure

P. 327, © Anthony Taber/Condé Nast, with permission from The New Yorker (Publications/www.cartoonbank.com); *p. 340,* Courtesy of Index Funds Advisors, Inc., Irvine, CA., Copyright 2010; *p. 342,* Courtesy of Psyleron, Inc.; *p. 345,* Courtesy of Index Funds Advisors, Inc., Irvine, CA., Copyright 2010; *p. 350,* free use; *p. 352,* Edward Hicks (1846) <http://commons.wikimedia.org/wiki/File:Noahs_Ark.jpg>

EPILOGUE:
The PEAR Experience

P. 365, Robert Jahn and Brenda Dunne; *p. 366,* Robert Jahn and Brenda Dunne; *p. 367,* Robert Jahn and Brenda Dunne

ACKNOWLEDGMENTS

The composition and technical production of this book could not have been accomplished without the astute and dedicated collaboration of many supportive and skilled friends and colleagues. In particular, we humbly doff our caps to two members of our PEAR/ICRL family who, despite woefully inadequate professional compensation, shouldered major practical burdens of its development. Our highly capable PEAR factotum, Elissa Hoeger, brought her extensive editorial skills to the preparation of a multitude of chapter drafts and their supporting references and illustrations. Elissa's self-effacing talents thread throughout the text, to her great credit and our great benefit. A similar generosity of technical acumen characterized the efforts of John Valentino to convey the essence of our research results via the long sequence of graphical illustrations that support the literary descriptions.

A few respected colleagues kindly reviewed an early draft of the manuscript and provided many suggestions and comments, most of which were incorporated into the final version. We are particularly indebted to Henry Bauer, Richard Blasband, Larry Dossey, Patrick Huyghe, and Teresa Wiater, all of whom drew from their particular professional and personal perspectives to make the final text undeniably richer.

For a variety of reasons, we decided to take advantage of the many recent advancements in self-publishing and print-on-demand technologies, rather than to engage a commercial publishing house. To this end, we established the ICRL Press, which has already produced a re-issue of *Margins of Reality* along with several other offerings. This strategy added to our team two experienced and talented colleagues: Patrick Huyghe, the highly skilled publisher of Anomalist Books, and Laura Smyth, a gifted book designer, both of whom were personally resonant with our goal of creating an appropriate and attractive presentation of this book, and to whom we owe immense appreciation and admiration. We are also especially grateful for a generous grant from the

BSW Foundation, honoring the memory of our late friend and long-time supporter of PEAR and ICRL, Sandra S. Wright, and providing much of the financial support for this book's publication.

Finally, we must reiterate our sincere gratitude to the many other members of the PEAR staff, its interns, and its precious operators who comprised the heart of the PEAR program and sailed so courageously with us on this long and arduous odyssey.

Robert G. Jahn
Brenda J. Dunne

INDEX

Index

Index

Mystical, 3, 21, 28, 30, 247, 273, 279, 303, 351
Mythos, 354

N

Near-death experience, 287
Nelson, Roger, 114, 357
Neurophysiology, 11, 16, 287, 290, 291, 293, 295, 331
Newton, Isaac, 28, 30, 322
Noah's Ark, x, 352
Non-linear dynamics, 90, 120, 187, 188
   chaotic systems, 38, 120, 187, 188, 258, 320
   complex systems, 38, 181
Non-locality, 10, 16, 17, 32, 35, 76, 142, 162, 165–166, 186, 189, 190, 233, 236, 237, 270, 329, 353, 354. *See also* Time independence, Distance independence

O

Objective/subjective. *See* Complementarity
OldREG experiment, 60
Old REG (OREG) replication experiment, 96
Operator characteristics, 36, 64, 68, 70–71, 179
   consistency of performance, 66–68
   gender, 64, 71, 80, 93, 127, 143, 168–174, 329
   multiple operators, 71, 78–82, 135, 203, 216, 217, 218, 329 *See also* Co-operators
   parameter preferences, 56, 57, 61, 64, 68, 69, 70–71, 86–87, 92, 127, 142, 143, 192, 259, 329
   signatures, 47, 53, 63, 68, 69, 70, 79, 126, 128, 328
   strategies, 19, 58, 59, 70–71, 89, 105, 111, 114, 127, 174, 201, 279, 283–284, 301, 307, 312
Optional stopping, 56

P

Paracelsus, 196, 240
Parapsychology, 30
Pasteur, Louis, 349
Pattern
   creation, 10, 156, 157, 163, 164, 182, 252, 298, 299, 341
   recognition, 182, 252, 341
Pauli, Wolfgang, 312
   "effect," 37
PEAR
   community, 351, 354, 366
   facilities, ix–x, 59
   closure, 129, 344, 365
   history, xii, 78
   purpose, xii, 34
   staff, ix, x, 58, 60, 119, 139, 207, 357
   website, xiii
   "PEAR Proposition" DVD/CD set, xiii, 345
PEAR 200 experiment, 80–82
Pendulum experiments, 139–147, 165, 168, 172
Péoc'h, René, 131–132
Percipient. *See* Remote perception
Physical currencies, 35–36, 264, 331, 332
Physiological sensors, 290–295, 298
Picasso, Pablo, 21
Placebo. *See* Medicine
Planck, Max, 263
Plato, 28
Play, ix, 21, 58, 122, 183, 189, 201, 335, 363
Poltergeists, 190
Popp, Fritz-A., 9
Practical Applications, xv, 15–19, 34, 38, 49, 189, 341–342
Prayer, 15, 16, 17, 30, 279, 286, 331
Precision vs. uncertainty, 19, 249, 254, 260, 281, 310–311, 321, 325
Precognition, 186, 331. *See also* Presentiment
Precognitive remote perception (PRP). *See* Remote perception
Presentiment, 247
Princeton Engineering Anomalies Research. *See* PEAR
Probabilistic analysis. *See* Statistical analysis
ProbREG experiment, 96–99, 100
Propitious environment. *See* Ambience

PRP. *See* Precognitive remote perception;
Remote perception
Protocols, (See also individual
experiments)
remote, 60, 76, 112, 142, 161
tripolar, 53, 55, 125, 128, 154, 166
parameters
automatic vs. manual, 57, 58,
68, 329
feedback modes. *See* Feedback
instructed vs. volitional, 57, 58,
68, 87, 88, 127, 142, 143, 145,
203, 216, 217, 219, 252
Pseudorandom experiments, 117–120
Psi, 68, 200, 255, 363
Psyche, 17, 37, 68, 156, 179, 237, 252, 254,
296, 354
Psychoneuroimmunology, 16. *See also*
Medicine
Psyleron, Inc., 342, 345

Q

Quantum mechanical model. *See* Models

R

Radin, Dean, 49
Random event generators (REGs), 23, 24,
49, 50, 52–64, 76, 106, 114, 123, 127,
128, 131, 162–166, 168, 171, 180–181,
279, 282, 328–330, 341–342
Benchmark experiments, 52–64, 60,
65, 66–76, 95–96, 179
co-operator experiments, 78–80,
135, 174
experimental units
Run, 57
Series, 57
Session, 57
Trial, 57
experiments, 48–50, 52–64, 78–10.
See also FieldREG
protocols, 55–60
pseudorandom, 117–120
remote experiments, 162–166
secondary parameters, 58, 68–70,
329
Random Mechanical Cascade (RMC),
122–129, 139, 147, 158, 165, 168, 172

Random physical processes, systems,
devices, xiii, 35, 42, 48, 50, 78, 105,
117, 155, 165, 284, 328, 354
Randomness, randomicity, xii, 7, 8, 9,
10, 21, 35, 105–109, 114, 117, 124, 128,
139, 147, 182, 183, 190, 245, 247, 248,
284, 325, 330, 340, 353
Reductionism, 16, 29, 30, 31, 317, 336,
339
REGs. *See* Random event generators
Reincarnation, 287. *See also* Survival
Religion, 29, 107, 321, 357
Remote human/machine experiments,
60, 76, 162–166
Remote perception, xiii, 161, 162, 196–
255, 311, 333
agent, 18, 160, 161, 199, 200–240,
242, 281
analytical judging, 204, 207, 214–215,
225, 251. *See also* Scoring
descriptors, x, 212, 215, 219,
220–221
artificial targets, 254
effect size, 216, 217, 218, 219, 221,
223, 227, 236, 241, 242, 243, 251,
252
empirical chance distributions, 213,
219
first trials, 222
implications and interpretations,
245–255, 281
participant attitude, 201, 259, 266,
298–299, 306, 312
percipient, 19, 160, 197, 200–240,
248, 250–255, 275, 281, 307, 312,
359, 360
multiple, 203, 216, 217–218
response biases, 219, 228
practical applications, 281, 282, 286,
288, 341, 343–344
protocols. *See* Scoring
scoring
ab initio, 207, 211, 212–217, 223,
225, 241, 242, 243, 251
binary, 211, 250
distributive, 208, 228–232, 243
ex post facto, 207–208, 211, 212,
213, 216–217, 227, 241, 243,
250
FIDO, 207, 208, 226–228, 231,
233, 242, 243, 251
ternary, 212

CPSIA information can be obtained at www.ICGtesting.com
Printed in the USA
BVOW08s0931240214

345829BV00008B/181/P

9 781936 033034